# STUDYING MST WITH THE OPEN UNIVERSITY: WRITING ASSIGNMENTS

BY

## DR. Peta Trigger Ph.D, Ed.D

Northampton Academy of Post Doctoral Studies, NN3 8TJ

kindle direct publishing

KDP

2 Emms Hill Barns

Hamsterly

County Durham

First Published 2015

(2nd Edition)

ISBN 978-1503332614

PRINTED BY CREATESPACE

https://www.createspace.com

# FOREWORD
## Studying with the OU

The author was a student at the Open University between 2000 and 2010 (11 years in all). She started with a 30 point level 1 Technology course, *You, Your Computer and the Net* and finished with two 30 point level 3 courses, *Electromagnetism* and *Quantum Mechanics*. The courses she took included six 10-point entry-level "Openings" (e.g. Y162 *Starting with Maths*) and "Breakthrough" courses (e.g. Y155 *Another Breakthrough to Mathematics, Science and Technology*), level 1 (S151) and level 2 (S292) Science 'Short Courses' and a mix of 30 and 60 point courses at levels 1 and 2, and all 30 point courses at level 3.

She graduated (from the OU) in 2009 with a B.Sc. in Mathematics and in 2010 with a B.Sc. (Open) in MST subjects, both with honours. Along the way, with 3-4 TMAs (Tutor Marked Assignments) for a 30 point course and 6-7 TMAs for a 60-point course, with some 720 points to her credit, she must have completed some 80 or so TMAs during her OU student career. (Many CMAs- Computer Marked Assignments and a number of ECAs (for those courses which had no examination) were also completed).

So, the author has had plenty of experience in writing OU MST assignments and the author feels herself qualified to pass on the benefits of that experience and learning to other students who are contemplating study with the OU, who are at the beginnings of studying for an OU qualification, and perhaps even those who have been studying with the OU for some time and are therefore more experienced students. After all, and this is also true in my case, we never stop learning and there is always room for improvement.

The author's approach to studying with the OU was to start with small beginnings, to 'try it out to see what it is like', with a level 1 Technology 30 point course with no exam at the end of it. Encouraged by the result and my ability to cope well with the intellectual requirements of level 1, the following year I decided to take the 30-point level 2 Technology course *Microprocessors* and the year after that a level 3 course *Radio Frequency Engineering*.

Up to this point, I had thought little about the qualification I might end up with. I was interested in Technology and Science, though, and moving forward in time to 2007, with 300 points under my belt, I was offered an open degree without honours by the OU. But I wanted an honours degree, and I found myself increasingly interested in Mathematics as it became more and more a part of study in going from level 1 to level 3 Technology courses.

It was then that I registered for the B.Sc. (Hons.) Mathematics (B36) qualification, and switched the focus of my OU studies (temporarily as it turned out) from Technology to Mathematics (and Statistics) courses. And having been 'bitten by the OU study bug', I began to take several courses in a year, in one or two years as much 100 points' worth.

Then, after completing my studies for the Mathematics degree, I realized that I could obtain another honours (Open) degree in MST, for which I needed just another 60 points at level 3 in one final year's study, having obtained 300 points from courses studied prior to the switch from Technology & Science to Mathematics. So I switched back from studying Mathematics courses to Science and

Technology courses, picking up (or finishing off) my initial studies with the OU.

In retrospect, it would have been wiser to accept the qualifications which are offered by the OU on the way to achieving an honours degree, beginning with that first open degree without honours. The Mathematics courses I took qualified me *either* for a degree in Mathematics *or* Mathematics and Statistics, and I could have obtained a Diploma in Statistics along the way.

So, my recommendation to others is that they start their study with the OU with an entry level 10 point or level 1 30 point course in a subject area which they are most interested in and (ideally) most knowledgeable about to give them a sound background for further study, and work up to courses at level 2 in subsequent years. Having successfully experienced enough study at level 2, level 3 courses can be taken in towards the end of their OU student career. I also recommend accepting all qualifications offered and for which one is eligible *en route* to one's final qualification(s).

Of course, course work in addition to the examination (or ECA) is an important element of assessment for any OU qualification. TMAs are normally a big part of course work (continuous) assessment, and to get a good degree, the student needs to put in a good performance in answering TMA questions. It is hoped that the advice in this book will help to achieve this.

# TABLE OF CONTENTS

|  | page |
|---|---|
| FOREWORD | 2 |
| LIST OF GRAPHS | 8 |
| LIST OF TABLES | 9 |
| LIST OF CHARTS | 10 |
| LIST OF DIAGRAMS | 11 |
| GENERAL INTRODUCTION | 12 |
| CHAPTER 1: TYPEWRITTEN AND HANDWRITTEN TMA SCRIPTS | 18 |
| CHAPTER 2: PRODUCING OU ENTRY-LEVEL MST ASSIGNMENT ANSWERS | 23 |
| APPENDICES TO CHAPTER 2 | 42 |

**CHAPTER 3: ENTRY-LEVEL MST MATHS TOPICS**     54

**CHAPTER 4: MATHEMATICAL DERIVATIONS AND PROOFS**     66

**CHAPTER 5: CONSTRUCTING EFFECTIVE MST GRAPHS AND TABLES**     70

**CHAPTER 6: OTHER CHARTS AND DIAGRAMS**     109

**CHAPTER 7: A REPORT-TYPE ASSIGNMENT QUESTION**     114

**CHAPTER 8: WRITING ARTICLES FOR THE WEB**     121

**CHAPTER 9: CITING REFERENCES IN ASSIGNMENT ANSWERS**     131

**CHAPTER 10: SUMMARY AND CONCLUSION** 158

..


**REFERENCES AND BIBLOGRAPHY** 171

# LIST OF GRAPHS

|  | PAGE |
|---|---|
| WALKING DISTANCE | 71 |
| WORKFORCE NUMBERS | 72 |
| SALES OF TWO BRANDS OF WASHING POWDER | 75 |
| POWER AND MAINS VOLTAGE VARIATION | 76 |
| TEMPERATURE CHANGE | 79 |
| TEMPERATURE | 80 |
| BIRTHS BY CAESARIAN SECTION | 82 |
| NUMBERS OF CHILDREN OWNING A TOY | 90 |
| PREDICTION FROM "         "        " | 91 |
| FORCE VS EXTENSION | 94 |
| STRESS VS STRAIN | 96 |
| HOMINOID BRAIN CAPACITY | 99 |

# LIST OF TABLES

|  | PAGE |
|---|---|
| NUMBER OF CHILDREN OWNING A TOY | 88 |
| THE CHANGING FACE OF STUDY AT THE OPEN UNIVERSITY | 101 |
| A RECIPE FOR OAT BISCUITS | 104 |
| STAGES OF STONE TOOL MANUFACTURE | 105-106 |
| QUANTITIES OF INGREDIENTS FOR OAT BISCUITS | 115 |
| AMOUNT OF FAT IN BISCUITS | 118 |

## LIST OF CHARTS

**PAGE**

SPECIES OF FISH CAUGHT IN AN AREA OF THE
NORTH SEA   [PIE CHART]                                73

MODE OF TRANSPORT [BAR  CHART]                         74

MEAN ELECTRICAL RESISTANCE OF RESISTORS IN 5
BATCHES (NON-ZERO ORIGIN ON Y-SCALE)[BAR
CHART]                                                 77

MEAN ELECTRICAL RESISTANCE RESISTORS IN 5
BATCHES (ZERO ORIGIN ON Y-SCALE)[BAR CHART]
                                                       77

BABIES DELIVERED BY CAESARIAN SECTION [BAR
CHART]                                                 81

A FOOD WEB PRIOR TO ONSET OF MYXOMATOSIS   109

SPECIES COLONIZING A ROCKY SHORE SITE      111

## LIST OF DIAGRAMS

**PAGE**

RECTANGULAR GARDEN WITH FLOWER BEDS         66

HUMAN BRAIN SHOWING LANGUAGE CENTRES        112

# GENERAL INTRODUCTION
## Why Use Supplementary Material?

Although course guides may offer advice on writing assignments, I sometimes found during my study with the OU that I needed advice which was both more general in its scope and more detailed in explication than that given in the course materials. I felt that I would benefit from reading 'model' assignment questions, supplied with course materials in some courses, but in addition with an explanation as to how their content, structure and layout were decided upon.

Where this is the case, supplementary material giving advice on how to plan and logically set down an argument; the use and structuring of paragraphs; and proper sentencing -since all of these aspects come into the effective communication of a mathematical argument, including those with a scientific or technology content, would be apposite. Nor need such material necessarily be directed towards MST writing ...there are certain general principles of good practice that can be applied to all writing.

I found the OU's *STUDENT TOOLKIT* (full references to the TOOLKIT series which are frequently referred to in this book are given in the section 'REFERENCES AND BIOGRAPHY' at the end of the book) entitled *Essay and Report writing skills* useful in this context. In writing assignments, for a high mark one should aim to include all relevant aspects and present them in logical order. Here, advice about *how to introduce* a piece of writing and what to include in introducing it; *how to arrange*, '*signpost*' and *link* the various sections in the body of the writing; *how to 'build' an argument*, give it *direction* and achieve *fluency*; is appropriate. Information given about what to write as *concluding remarks*, to show *how the piece of writing has answered the question* or demonstrated its purpose is also useful. Finally advice about 'editing' and 'proof-reading', and useful pointers as to *how to refine* a piece of writing and improve its presentation, is also apposite.

How to *construct* TMA answers is one issue of primary concern to students, another is *content*. Some students may feel they lack a sufficient grasp of the subject matter or topic of an assignment question and that they need extra help. Different texts may differ in the approach used to explain the same topics. Just as there are often several methods of arriving at a solution to a given type of mathematical problem, there are usually different ways of explaining the underlying concepts, each with its own merits and demerits. Some students may find one method more suited to their style of learning, other students an alternative method, or students may benefit from an explanation of the same topic from a different angle.

An alternative approach to getting ideas across, with different or more examples of the same type, can be beneficial in other ways, too. For example, an alternative approach may 'work' better for some topics and some students, and help to reinforce mathematical concepts in others. Apart from this, other sources may be read for the additional information they contain, to further develop one's grasp of mathematical concepts. This also applies to scientific and technological concepts. Though *Mathematics* has been specifically referred to here, ther above remarks apply to other types of subject matter in *Science* and *Technology*.

For some students, one reason for reading supplementary material is in order to overcome lack of confidence in certain topics, or in applying mathematics to scientific or technological problems. And to develop a point made previously, such students may also use suitable supplementary material as an additional source of information about these topics and of extra examples/questions for practice. For though I found courses often came with its own supplementary booklets containing extra examples for practice to help drive home a topic, for some students and some topics, those supplied may not be enough. There may be too few questions about some aspects of some topics even considering the main text and the supplementary material together, to make the ideas to 'sink in'. Again, this also applies to scientific and technological topics and concepts as well as mathematical ones.

This book, then, gives advice on and examples of some of the main skills involved in writing MST assignments. To this end, assignments and articles written by the author when she was a student studying with the OU are both featured and drawn upon and extended. These deal individually with such topics as appropriate choice of medium, planning, structuring and writing TMA answers to essay and mathematical assignment questions, understanding and use of basic mathematical concepts, how to derive a formula and construct a mathematical proof; how to construct effective MST graphs, tables, charts and other diagrams; how to write answers to more specialized forms of MST assignment questions involving the preparation of a report and writing an article for the Web; and finally how to cite references.

Individual chapter conclusions containing a summary are provided where a previously written article forms the basis of the chapter, and in addition an overall summary is provided in the last chapter of the book. Where appropriate, appendices and references appear at the end of certain chapters for the same reason; the remaining references, as is usual practice, are given in a list in a separate section following the final chapter.

Although the articles drawn on to write this book were written with a view in mind to their application to specific courses which the author was studying at that time, the material is quite general and not specific or confined to any particular course or type of assignment. Obviously, though, any advice given would be too theoretical and difficult to follow without the benefit of illustrative examples and of limited practical use to students in writing answers to the particular assignment questions they are faced with. Given that illustrative assignment questions with appropriate content had to be used in order to explain the method, the aim in choosing them was that they be realistic but simple enough for most MST students including beginners to comprehend, so that students could focus away from the particular content of assignment question and concentrate on the strategy used in producing an answer to MST assignment questions.

So, what is important to readers in considering the steps in *construction* of an answer to a MST assignment question is not the particular courses to which they relate, or specific topics and content of the examples given, but the approach, methods and skills used on them.

However, the basics of TMA construction having been thus explicated and exemplified, following on from this, more typical examples in the author's answers to assignment questions which she tackled during her study with the OU are discussed to throw further light on the process.

Before s/he can begin to write an answer to a MST assignment question, the student must decide on the most appropriate medium in which to present it, be this in the form of handwritten text, typed or word-processed text, or a combination of the two. The opening chapter of this book considers these methods and their role in producing a clear and legible piece of writing, and an example of the consequences when the outcome is not.

# CHAPTER 1: TYPEWRITTEN AND HANDWRITTEN TMA SCRIPTS

## A Warning About Handwriting

Clearly legible handwriting is a must for assignments (not to mention exams). Below is shown part an assignment for MU120 which shows the possible consequences answers are difficult to read:

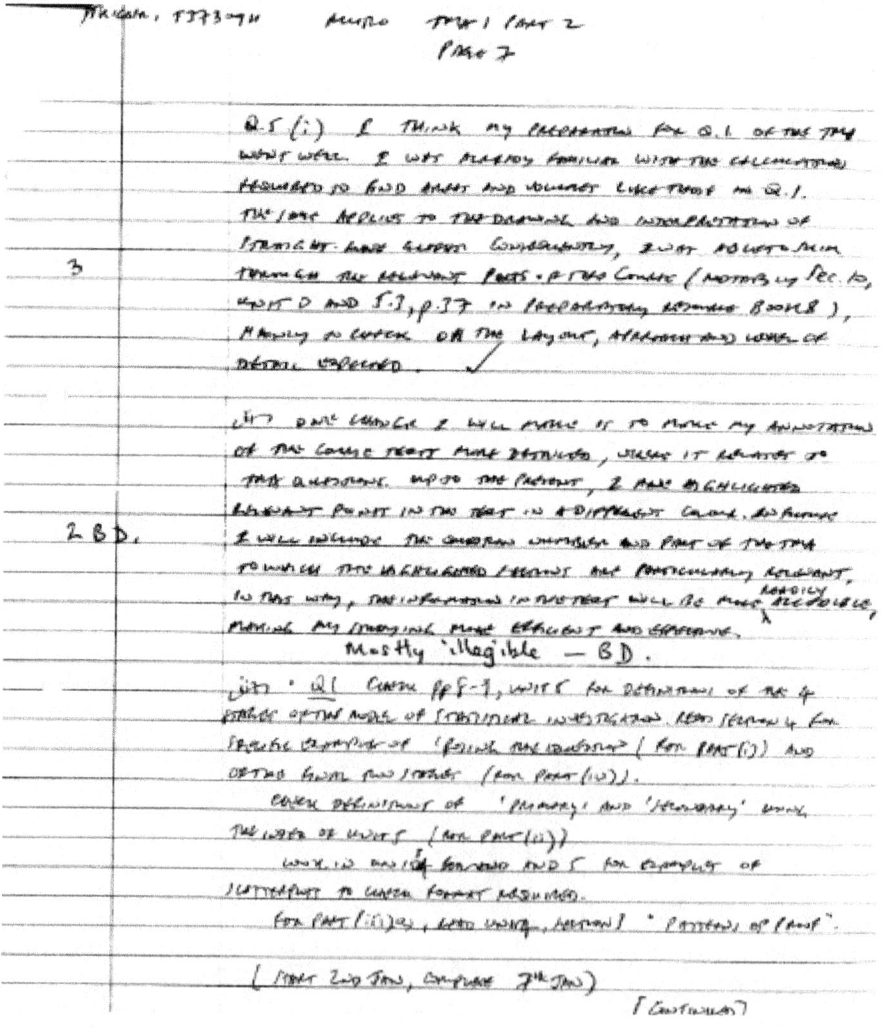

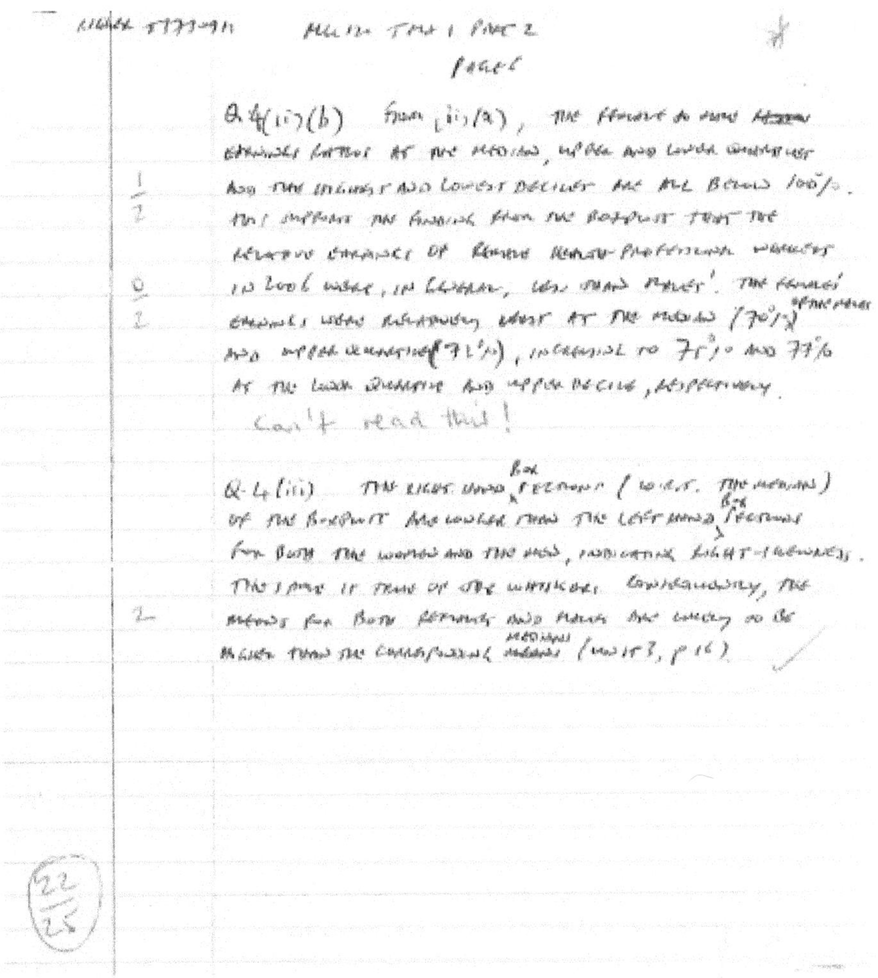


Naturally, if the tutor can't read the student's script, s/he cannot expect to receive marks for an answer even though it is correct. In the above case, this has resulted in a loss of 3 marks for this one part of one question alone, and this TMA contained 25 parts in all.

Notice that it was written in capitals to try to increase the legibility of poor handwriting. Aside from the fact that this tutor said *he* (still) couldn't read it, writing in capitals is slower. His recommendation was to "word process" future assignments (see front cover), but typing in all mathematical symbols using a standard qwerty keyboard causes extra difficulty and it is also more time-consuming and tedious. A solution was to combine the two methods- typing in the words and writing in the symbols by hand:

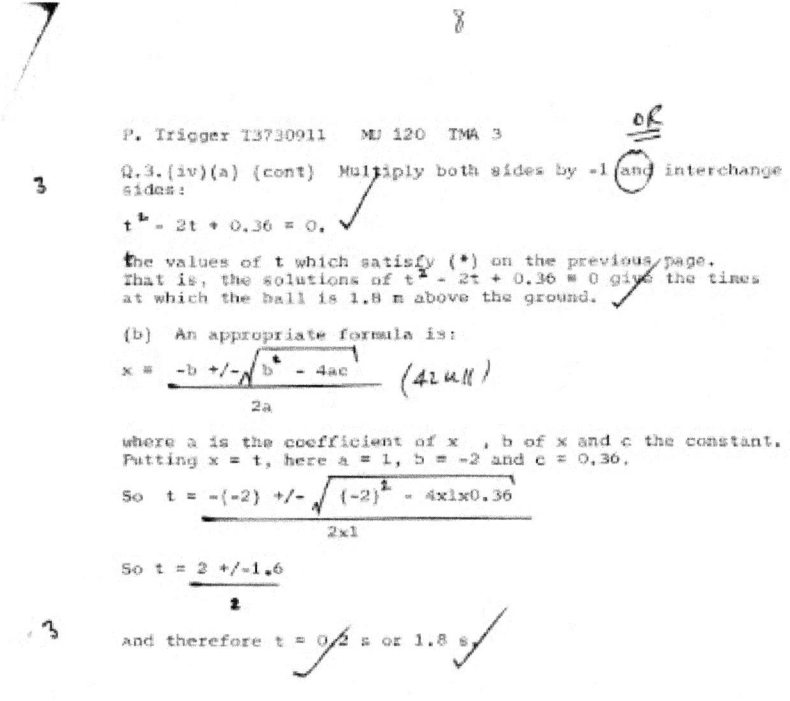

Above, a hand-drawn square root symbol has been added to the typescript. The *main thing* is any piece of writing should be clearly laid out and *legible*. It would have been neater to use appropriate software to produce graphs, but since legibility is more important than neatness, where the question requires it they can be hand written and included with the typed assignment, as shown below in parts of two assignments, one for MU120 (above) and another for SMT 359:

Having considered methods of producing the text of assignments, we move on in Chapter 2 to discuss in detail their planning, structuring and layout.

# CHAPTER 2: PRODUCING OU ENTRY-LEVEL MST ASSIGNMENT ANSWERS

## INTRODUCTION

The OU's 'Openings: A Guide to Student Services' (2004), ['Openings' courses were entry-level 10-point courses designed as 'taster' courses -*i.e.* as an introduction to undergraduate study] recommends that students read supplementary material to sharpen their study skills.

In the following paragraphs, I describe, in appropriate detail, one approach to writing answers to assignment questions, with particular reference to an example of an essay type question[1] (Superscripts used here refer to NOTES on p. 41). As noted in the General Introduction, the reader is reminded that the specific content of the question itself is not important here, but rather the general method of tackling such questions, which it is the aim of this chapter to explain. The applicability of the advice given to a question consisting of a mathematical problem is also discussed.

## THE SCOPE OF THE ADVICE GIVEN

To remind the reader, the advice given here is not directed towards writing assignments for any particular course specifically, though examples from the courses studied by the author will be discussed after expounding the method and illustrating it in application to some more basic examples, and in fact advocates the use of general principles of good practice that apply to the writing of MST assignment answers in general.

## A FRAMEWORK FOR WRITING ASSIGNMENT ANSWERS

The fundamental 'principles of good practice' in assignment writing can be identified with a sequence of STAGES to be worked through in producing an effective answer to an assignment question. Described here in detail, these stages provide a procedure for determining what content to include and how to structure it. In the authors' view, even short pieces of writing (which is the case with the initial example assignment questions discussed here) benefit from this approach. Also important are proper sentencing and paragraphing.

Overall, then, the 'principles of good practice' referred to fall into two main categories. The first concerns the use of a framework for tackling assignment questions using a sequence of stages. The second is concerned with both content- deciding on what goes into the answer in terms of the ideas included- and with how these ideas are structured and organised. In the BROAD STRUCTURE of the answer, an *introduction*, a *main body of argument, explanation or description* and a *conclusion* should be discernible. The FINE STRUCTURE relates to the sequencing and linking of ideas in paragraphs and sentences, to produce a logically organised, coherent and fluent answer.

Beginning with the stages of Understanding the Question, Producing a Plan and Writing the Drafts, these principles of good practice will be discussed here in turn, and examples from actual assignments as completed by the author will be given. In doing so, material from several of the booklets in the OU's *TOOLKIT* series will be drawn upon as appropriate. The following example assignment question will be used for illustrative purposes:

> "In studying 'Exploring Pattern' [an *OPENINGS* course book], write an account of one topic which you found easy and one which you found more difficult. Which strategies did you use or might you have used in coping with the more difficult topic (max 300 words)?"

The process of answering this question will be spelt out here in detail for the purposes of explanation and illustration. This might give the impression that much more work is involved in producing such an answer than is actually the case.

## STAGES OF WRITING AN ASSIGNMENT ANSWER

### (1) UNDERSTANDING THE QUESTION

The initial stage is to understand the requirements of the question, i.e., WHAT the question is asking (the content required in the answer) and HOW it is to be answered (the approach required). In an assignment question, there tends to be *key words or phrases* which define these requirements.

In the example question, key words/phrases are 'Exploring Pattern', 'account','one topic', 'found easy', 'found difficult' and 'strategies'. Identifying such words and accurately discerning their meanings determines the scope of the question and hence the content and approach of the answer.

### (2) PLANNING AN ASSIGNMENT ANSWER

In the present example, 'content' words ('Exploring Pattern', 'topic') help to make the content required clear (though the particular topics 'found easy' and 'found difficult' will vary from student to student). 'Process' words- 'account' and 'strategies'- require more careful thought in order to get the approach right in the answer. Here, one of the OU's *Toolkit* booklets was used to obtain a precise definition of 'account' as "an explanation with reasons" (an account is therefore not merely a description as might be supposed). 'Strategy' was looked up in a dictionary, which defined it as a plan of action.

Having determined the scope of the question, the next stage is to write a plan of the answer. This may be usefully conceived as a three step process:

>(1) **'brainstorming'** ideas,

>(2) **selecting and sequencing** ideas

and

>(3) **structuring, allocating and organizing ideas** under headings of 'Introduction', 'Main Body of Argument, Explanation or Description' and 'Conclusion (these headings do not have to appear in the answer itself, though 'Introduction' and 'Conclusion' might be useful to head the associated sections). Often, a Summary section or a summing up of what has been said is appropriate at the end, either separately or as part of a 'Summary and Conclusion(s)' section.

## BRAINSTORMING

The purpose of this activity is to generate any ideas which come to mind for possible inclusion in the answer, in any order. These can be single words, short phrases or sentences. Some of these ideas may turn out to be unsuitable for inclusion in the answer (for example as not sufficiently relevant, perhaps superfluous or redundant). However, with one eye on the question and the other on any word limit$^2$, the aim is to produce a pool of information containing a sufficient number of ideas relevant to the TMA question to include in the answer.

Guided by the key words in the example question, I wrote down several topics which I found 'easy' and several more difficult topics. Next I noted the reasons why I found them so. I also tried to pin down precisely which aspects of the topics I found easy or difficult. The nature of these difficulties gave clues as to which strategies as asked in the question were, or might be, suitable in coping with them. Additionally, The *Sciences Good Study Guide* (SGSG) was used as a resource of coping strategies (see the first section of the PLAN in Appendix 1a below).

**SELECTING AND SEQUENCING**

From the list of topics generated, I was able to select two topics for discussion in the answer. 'Graphs' was chosen as the 'easy' topic and 'transformations' as the difficult topic. These particular topics were chosen firstly because they are fundamental and secondly because they generated plenty of ideas. These then became the subsequent focus of my plan.

After rejecting irrelevant and unimportant ideas, those remaining had to be arranged in order. This was done mainly in two ways:

> (1) by reflecting on those content or study aspects of the chosen topics which were central to the ease or difficulty with which they were tackled, and those which were less so. So, for example, the working of sufficient examples involving graphs was seen as more central to a grasp of the topic than learning the 'language' used to describe graphs. Therefore, the former aspect was placed before the latter;

> (2) by considering possible interdependencies and interrelationships between these aspects. Thus, consequential aspects were placed so as to follow their antecedents. For example, an inappropriate allocation of study time was a consequence of overconfidence, and so these two aspects were ordered accordingly.

***STRUCTURING IDEAS*** *FOR THE THREE MAIN SECTIONS OF THE ANSWER*

An essay type assignment question should contain definite *beginning*, *middle* and *end* sections. Each of these sections has a distinct purpose in giving the answer a sense of direction and in guiding the reader through it.

***ALLOCATING AND ORGANIZING THE IDEAS IN EACH SECTION***

Having brainstormed and sequenced the ideas to be included in the answer, the next step in planning was to determine which ideas to include in each section. The organisation of ideas within each section was based on the outcome of the sequencing process discussed above. Some ideas for fine structuring the answer were also written in the plan (see Appendix 1a).

The fleshing out of ideas in the answer was guided by the definitions of 'introduction', 'main body of argument, explanation or description' and 'conclusion' referred to in the booklets. The broad structuring of an answer in this way is described under the next heading.

## (3) WRITING A DRAFT OF THE ANSWER

### BROAD STRUCTURE

The *introduction* serves to lead the reader into the main body of argument, explanation or description in the answer. Its principal purpose is to indicate how the main body of argument, explanation or description in the answer will be laid out, referring to the content and the approach that will be used. Clearly, the introduction, and indeed the answer as a whole, should be *based entirely on the requirements of the question* and, most appropriately, *in the order in which they are set out there* (see para. 1 of Appendix 1b).

The *main body* of explanation (in this case) in my answer contains the detail of the discussion of the two topics found easy or difficult, and the reasons why they were found so. Strategies for coping with the 'difficult' topic follow from these reasons. This section of my answer to the example question is shown in Appendix 1b, paras. 2-4.

Finally, the *conclusion* sums up the points made in the main body of argument, showing how the requirements of the question have been met.

## FINE STRUCTURE

### PARAGRAPHING

I arranged the ideas generated in the plan in paragraphs using the advice given in the *Student Toolkit* booklet 'The Effective Use of English'. The order and emphasis of the requirements set out in the question provided a framework for sequencing the paragraphs in the answer. Hence, following a paragraph of introduction, a short paragraph about the 'easy' topic and then a somewhat longer paragraph about the 'difficult' topic, were decided upon. *The longest paragraph* about 'coping strategies' followed, *reflecting its importance in* the context of *the question*. The answer finished with a concluding paragraph. This scheme also met the requirement of '*one idea per paragraph*' mentioned in the booklets.

*In writing each paragraph*, I aimed to *make its subject clear in the opening sentence*. This subject could then be expanded upon in the remaining sentences of the paragraph. For example, the first sentence in the second paragraph in the answer (see Appendix 1b) makes clear that the paragraph is about 'graphs' and why I found this topic easy.

The various paragraphs were *linked* together using 'transitional' words and phrases. For example, having discussed the 'easy' topic, it seemed appropriate to introduce the following paragraph about the 'difficult' topic with the phrase, 'In contrast...'. Paragraphs 4 and 3 were linked by showing how the content of paragraph 4 follows on from paragraph 3. There are several other examples to be found in the answer, which readers may care to try and spot.

**SENTENCING**

Other such devices were used to link sentences within paragraphs. Examples are the use of 'Hence', 'Likewise' and 'as a consequence' in paragraph 2, and the use of 'first' and 'second' in paragraph 4. Further examples may be seen in the answer, which again readers should identify for themselves.

I have used *'specialist' words* as appropriate and expected in TMA answers in referring to topic content. 'Transformations', 'reflections' and 'rotations' are examples. Though some *sentences* are long, I have tried to achieve a balance and *introduce variety*, not only *in length*, but in *wording, phrasing, emphasis, punctuation* and in the use of *parenthesis,* which all improve interest and readability.

**4. SUBSEQUENT DRAFTS**

The finished answer is a result of several subsequent drafts. Essentially, subsequent drafting means going over the previous draft to improve it. In answering the example assignment question, this involved changing the order of presentation of some ideas to improve the logical progression and fluency of the answer. Also, the lengths of some sentences had to be reduced to meet the imposed word count[2], whilst still effectively communicating the main ideas. To achieve this purpose, several sentences containing less important or redundant ideas had to be omitted altogether.

## 5. CHECKING THROUGH THE ANSWER

After writing the title, Appendices and References (usually required: Citing References is the subject of Chapter 9 in this book) a penultimate read through checked for grammatical errors and spelling. Finally, the answer was re-read, with the question fully in view, to check that it sounded right and fulfilled all the requirements of the question. It is very beneficial to put an answer to one side for a week or two, before returning to scrutinize it afresh, when further improvements might well become apparent. Seeking the alternative points of view, comments, advice and criticisms of others (e.g. fellow students) invited to read the script could also be very helpful.

The completed answer is in Appendix 1b.

**A further example in answer to a 'biological' essay-type assignment answer is in Appendix 1c.**

## WRITING ANSWERS TO MATHEMATICAL PROBLEM-TYPE ASSIGNMENT QUESTIONS

It remains to consider how to apply the above advice in answering *mathematical problem-type questions*. The advice given so far has been given almost exclusively with reference to essay type questions. Hence, it might seem that the applicability of the advice given might depend on the extent to which an effective mathematical problem-type answer contains essay-type features. This issue will now be considered under the headings previously used in discussing essay-type answers. For the purposes of illustration and discussion, the following question is deliberately chosen for the simplicity of its conceptual context so as to avoid distraction from any possible *mathematical* complexities involved, since again it is the general *method* or approach to answering mathematical problem-type assignment questions that is the focus here, not its specific content:

> "The perimeter of a rectangle is 30 cm. If the length of the rectangle is 5 cm, find the width of the rectangle."

**THE STAGES OF ASSIGNMENT ANSWER WRITING**

The initial stage of **understanding the question**, and the advice given to identify the key words in the question and carefully discern their meanings, seem as relevant in this case as in essay-type questions. Clearly, it is important to identify and correctly interpret the key (content) words 'perimeter', 'length', 'width' and the numerical values given in the question, as well as the process word 'find'.

In regard to the **planning** stage, a detailed plan of a short, 'closed'[3] mathematical answer involving the single application of a simple formula is obviously unnecessary. One preliminary draft followed by a second to tidy up the layout would probably be all that is required.

But detailed planning might well be advisable where a question is more open-ended or where, for instance, an answer requires the application of several, perhaps more involved, mathematical techniques with more steps in the working. This being so, it is likely that a more comprehensive drafting stage similar to that described for essay-type answers would be appropriate in setting out, explaining and refining the argument. The associated advice given for producing answers to essay-type questions would therefore be more useful in this case.

# Writing OU MST Assignments

**BROAD STRUCTURE**

The example given here suggests that, in terms of broad structure (introduction, main body of *argument* (in this case) and conclusion), both types of (essay and mathematical) answer share the same features. A suitable answer to the above question might be *introduced* by quoting a formula for the perimeter of a rectangle. I.e.,

> "The perimeter P of a rectangle is given by:
>
> $P=2L+2W$,"
>
> where L and W are respectively the length and width of the rectangle."

Hence, as in the essay-type answer, an introduction indicates what the content of the answer will be and the approach that the answer will take.

The next section of my answer is also familiar as the '*main body of argument*, explanation or description' found in essay-type answers:

> "L is given as 5 cm.
>
> P is given as 30 cm.
>
> Rearranging the expression for P to make W the subject:
>
> $P-2L=2W$

and so $W = \frac{1}{2}(P - 2L)$.

Substituting the values of P and L,

$W = \frac{1}{2}(30 - 2 \times 5)$ cm

$= \frac{1}{2} \times 20$ cm

$= 10$ cm."

Finally, it is appropriate to *conclude* the answer by stating the result of the calculation in terms of the question put:

"Therefore, the width of the rectangle is 10 cm.".

Again, then, as with the essay-type answer, the answer finishes by showing how the requirements of the question have been met.

The complete answer is shown in Appendix 2.

### FINE STRUCTURE

As can be seen from this example, even in such a short and simple answer, mathematical (i.e. symbolic) statements are accompanied by verbal statements which 'signpost' (*SGSG*) the argument, making it more accessible to the reader. It was noted earlier that essay-type answers benefit from appropriate paragraphing and sentencing, which guide the reader stepwise through a logically ordered argument. The fluency of an argument is also thereby improved. Appropriate paragraphing and sentencing is just as important in mathematical problem-type answers, perhaps more so, since otherwise the communication of the argument set out in the answer might become too dependent on mathematical symbolism.

....

## CONCLUSION

In this chapter, in writing entry-level MST assignment answers, the advice given was seen as falling into two main categories:

(1) providing a framework for tackling assignment questions using a sequence of stages. These are fundamentally the stages of understanding the question, producing a plan, and drafting the answer;

and

(2)
(a) supplying a broad structure for the answer in terms of an introduction, main body of argument, explanation or description, and a conclusion;

(b) providing guidance on fine structuring an assignment answer via appropriate paragraphing and sentencing. This advice relates to how the paragraphs and sentences used might be constructed, ordered and linked.

Although the advice given in the *STUDENT TOOLKIT* booklets is not specifically directed to any particular course, I have shown in detail, using an example, how the advice given can be applied to MST essay-type assignment questions in general. A further example is given in Appendix 1.

In addition, it was shown by reference to another example that the advice given in Category (1) above about answering the question, is as applicable to mathematical problem-type questions as it is to essay-type questions. The finished product is in Appendix 2.

Though some planning and drafting would be desirable, detailed planning and drafting of the kind described for an essay-type answer would not be required in the case of short, closed mathematical problem-type questions. However, it was argued that this might well be appropriate in tackling questions of a more open or involved nature.

Finally, it was shown that effective mathematical problem-type answers and essay-type answers share broad and fine structural features. This suggested that the advice given in Category (2) above in relation to paragraphing and sentencing is also be useful in producing effective answers to mathematical problem-type questions.

.....

---

**NOTES**

[1] The study experiences referred to in answering this question are those of a fictitious student.

[2] Check with your tutor for any tolerance allowed in the word count. I recommend that the full word count be used. If your answer falls short of this by more than 10%, it is an indication that not all the required information has been included. Often, tutors allow 10% over the word limit before deducting marks for producing an assignment which is too long.

[3] In closed questions the steps of working are prescribed, implicitly at least, by the question.

# APPENDICES TO CHAPTER 2

## APPENDIX 1a: STAGES IN PRODUCING AN ANSWER TO THE ESSAY-TYPE QUESTION DISCUSSED IN THE TEXT

QUESTION:

"In studying 'Exploring Pattern', write an account of one topic which you found easy and one which you found more difficult. Which strategies did you use or might you have used in coping with the more difficult topic
(max 300 words)?"

### PLAN

EASY TOPIC

DRAWING GRAPHS×
Reasons:
Well practised.
Understand the algebra and language.
Liking for, confident and quick in doing.

Use of Formulae
Reasons:
Straightfoward substitution of numbers in place of symbols.


Area/perimeter
Reasons:
Familiar opics

DIFFICULT TOPIC

Spirolaterals
Unfamiliar language
Unsuccessful diagrams



TRANSFORMATIONS×

Reasons:
Overconfidence in:
'Reflection', 'Rotation'
Poor concentration.
Ineffective use of study time.

Scatterplot
Reasons:
Unfamiliar topic

# Writing OU MST Assignments

***BROAD STRUCTURE***

**INTRODUCTION**

Topics× chosen fundamental.
Reasons for ease/difficulty are basis for 'coping strategies'.

**MAIN BODY**

(1) Description of 'easy' and 'difficult' topic content.
(2) Study problems.

GRAPHS:
Content:

Interpreting/drawing.
$y=mx+c$.
Scaling axes; plotting points, reading from.

Reasons found easy:
Practised. Confident. Understand key ideas.

TRANSFORMATIONS:
Content:

Reflections/rotations. Translations ok.
Example from Module1.

Reasons found difficult:
Overconfidence in 'basics'. Mistakes in applying to patterns.

Concentration affected-attempted too much/session.


COPING STRATEGIES;

Careful study of definitions. Work more examples for understanding/skill in applying.

Break down concepts, study in smaller 'chunks'. Share difficulties with tutor/other students.

### CONCLUSION
- Sum up: main point is to ensure understanding of underlying ideas.

- Use problems encountered as basis for coping strategies.

### *FINE STRUCTURE*

Main paragraph order (follows order of question): easy topic, difficult topic, then coping strategies.

Links between paragraphs: contrast easy and difficult topics. Coping strategies based on specific difficulties.

Sentence links: interrelate reasons.

The completed answer is shown below:

# APPENDIX 1b: THE FINISHED ANSWER TO THE ESSAY-TYPE QUESTION DISCUSSED IN THE TEXT

QUESTION:

"In studying 'Exploring Pattern', write an account of one topic which you found easy and one which you found more difficult. Which strategies did you use or might you have used in coping with the more difficult topic
(max 300 words)?"

ANSWER:

"Below I discuss one major topic in 'Exploring Pattern' which I found easy and another found more difficult, with reasons. Based on these reasons, I describe some strategies for coping with the latter.

I found 'graphs' easy (e.g. Ch. 2), having previously worked many examples involving drawing and interpretation. Hence, I was already skilled in deciding scales for the axes, plotting points and reading data from graphs. Likewise, I was familiar with the 'language' of graphs and graphs as a representation of '$y=mx+c$'. Consequently, I felt confident in tackling graphical problems.

In contrast, I encountered problems with 'transformations' (Ch. 4), essentially through overconfidence in my grasp of 'reflections' and 'rotations'. As a result, I spent too little time on these "basics". I then found it difficult to concentrate on more complex symmetric patterns. So, for example, in the Ch. 4 activities, I made mistakes in identifying the centre and amount of rotation of base tiles required to build up the patterns.

From the above difficulties, with help from SGSG, I identified several coping strategies. The first was to look carefully at definitions of the key concepts of 'rotation' and 'reflection'. I then worked through sufficient examples to achieve understanding and skill in applying them to symmetric patterns. The second, related to poor concentration, was to allocate more study time to this topic in the Module, whilst limiting the scope of each session to one sub-concept (e.g. rotations: amount of turning). Looking back, tutor advice and discussion with other students might have been beneficial.

Summing up, 'graphs' was easy for me because I understood the key concepts and had developed the necessary skills through practice. The opposite was true of 'transformations', and this led to problems in concentration and of inadequate study time. However, coping strategies based on these difficulties enabled me ultimately to master this topic."

## APPENDIX 1c: PRODUCING AN ANSWER TO "BIOLOGICAL" SCIENCE ASSIGNMENT QUESTION ESSAY-TYPE QUESTION (level 1, Y153)

QUESTION:

> "Write an account explaining the changes in the number of buzzards and red deer after myxomatosis, using appropriate technical language. (Maximum 400 words)."

**COMMENTS**

**Understanding the Question**

Before writing this assignment, I underlined in red each of the key words and phrases in the question. These were <u>changes in numbers of</u>; <u>buzzards</u>, <u>deer</u>; <u>myxomatosis</u>; <u>technical language</u>.

This was done so as to be sure of centring the piece on these key phrases and words. So, for example, instead of referring to deer as being 'animals who eat only vegetable matter', the technical word "herbivore" was used. "Myxomatosis" is a technical term specifically mentioned in and a focal point of the question, and so should be briefly defined in the answer (para. 2)

**Brainstorming and PLAN**

I wrote down all the ideas I could think of for possible inclusion in the answer, guided by the key phrases and words of the question, and then structured and arranged them as follows:

INTRODUCTION- *myxomatosis* -*available food*- deer & buzzards

*myxomatosis*- definition, transmission & effects

**deer** *herbivore* population-rabbit numbers decrease> more grass, plants, tree saplings, etc. for deer> deer numbers increase

**buzzards**- *carnivore* prey on rabbits, insects, birds - numbers decrease because fewer rabbits to eat. (Also) some insect larvae eg minotaur beetle feed on rabbit dung; some birds eg wheatears fewer rabbit holes for nests.

CONCLUSION-myxomatosis>rabbits decrease>more vegetation for deer>deer increase; rabbit prey for buzzards>buzzards decrease.

.

The completed answer to this assignment question is shown below:

# APPENDIX 1d: THE FINISHED ANSWER TO THE "BIOLOGICAL" SCIENCE ASSIGNMENT ESSAY-TYPE QUESTION

P. Trigger I3730911   Y153 TMA 02                    TMA 02

Q.2(c)

After myxomatosis, more food becomes available for red deer but less for buzzards. How this happens and how numbers of red deer and buzzards are changed as a result, are explained below.

The myxoma virus is transmitted via fleas, mainly from dead to living host rabbits. This disease kills rabbits. (p. 60, *Environment*).

As a result red deer numbers increase. This happens when the amount, height and type of local vegetation change due to falling rabbit numbers. Like red deer, a competing herbivore, rabbits eat grasses and leaves. Dense rabbit populations keep grass short and may cause serious erosion problems (p. 62, *Environment*). Hence, fewer rabbits mean more grass for red deer to eat, supporting greater numbers.

Other plants are similarly affected. Rabbits like bushes such as juniper and hawthorn. These thrive when eaten by fewer rabbits. Fewer tree seedlings are killed by rabbits, allowing trees to grow. This increase in bush and tree growth provides more food for red deer, sustaining increased numbers.

Taller plants spring up and take over by shading established species. Reduced biodiversity may result (Table 3, p. 63, *op. cit.*), but the overall increase in food available to red deer supports more individuals.

Edible species replace unpalatable species such as elder when fewer rabbits eat competing plants. Scratching of bare patches by rabbits lessens, restricting the habitat for tough species such as ragwort.

In contrast to red deer, buzzard numbers decline after myxomatosis. A major reason is that rabbits are their main food. As consumers, buzzards cannot synthesize

> Q.2(c) [cont]
>
> nutrients. As carnivores, neither can they turn to plants. However, raptors can take mice, birds and large insects (see Q.2(b) Fig. 1, p. 3). But these may not make up for the rabbit shortfall. More of smaller mice prey would have to be caught, for example.
>
> Rabbit numbers also affect bird and insect species. Lessening rabbit activity spells a decline in habitat for some species. For example, wheatears nest in holes left by rabbits. Fewer rabbits mean fewer holes and fewer wheatears. Populations of bird species which nest in short grass, such as the stone curlew, decrease.
>
> Moreover, insect prey may wane with the declining rabbit population. For example, minotaur beetle larvae feed on rabbit dung. Fewer rabbits mean fewer larvae survive to become adult beetles.
>
> ...
>
> So, after myxomatosis, decreasing rabbit numbers allow the growth of grass, shrubs and trees to support increasing numbers of deer. But buzzards are dependent on rabbit prey and therefore the numbers of buzzards decrease.

Notice also how this essay has a definite Introduction (para. 1), Main Body of Argument, Explanation and Description, and a Conclusion (last paragraph). After the introduction, the subsequent block of paragraphs is addressed to explaining changes in red deer, followed by a second block of paragraphs explaining the changes in buzzard numbers.

After writing an initial draft, the essay was checked for comprehensiveness and logical progression.

Almost inevitably, if all relevant information is included in the answer, the first draft will be longer than that typical of OU TMA maxima- in this case the word limit is 400. So, the 2nd draft aimed to excise all inessential words, expressing meanings with few words, and ultimately, paring of the least important points to reduce the finished product to 400 words.

# APPENDIX 2: THE ANSWER TO THE MATHEMATICAL PROBLEM-TYPE QUESTION

**QUESTION:**

"The perimeter of a rectangle is 30 cm. If the length of the rectangle is 5 cm, find the width of the rectangle."

ANSWER:

"The perimeter P of a rectangle is given by:

$P = 2L + 2W$,

where L and W are respectively the length and width of the rectangle.

L is given as 5 cm.

P is given as 30 cm.

Rearranging the expression for P to make W the subject:

$P - 2L = 2W$

and so $W = \tfrac{1}{2}(P - 2L)$.

Substituting the values of P and L,

$W = \tfrac{1}{2}(30 - 2 \times 5)$ cm

$= \tfrac{1}{2} \times 20$ cm

$= 10$ cm.

Therefore, the width of the rectangle is 10 cm."

# Writing OU MST Assignments

Chapters 1 and 2 have considered the basics of TMA answer production in terms of choice of written medium, planning, structuring and lay out.

Chapter 3 will deal with some common basic *mathematical* ideas, an understanding and application of which may be an essential part of getting a good mark for answers to entry-level MST assignments. But again in principle the general advice given in relation to the use of supplementary material and mathematical ideas is also applicable to the *scientific* and *technological* ideas which figure in MST TMAs. Chapter 4 gives an example of the derivation of a mathematical formula which may also be asked for in a mathematics-based part of an MST assignment question.

# CHAPTER 3: ENTRY-LEVEL MST ASSIGNMENT MATHS TOPICS

**INTRODUCTION**

There are several reasons as to why students of MST subjects might find the following advice useful. It represents an additional source of help; it offers further examples on which to practise mathematical skills, and it helps students to increase their confidence and ability in number work and algebra necessary in tackling many MST assignment questions.

In addition, as noted previously, sometimes a slightly different approach to explaining an idea or the working of examples provided in a different text succeeds better in getting an idea across to some students.

One purpose of the following paragraphs is to supplement the material in courses and *SGSG*. A second purpose is to compare the approach with course-work SGSG approaches, where they may differ.

What follows also contains advice about analysing course content and TMA questions in terms of mathematical topics. Armed with such an analysis, students may then decide which topics and more specifically which concepts

# Writing OU MST Assignments

(if any) they need help with.

In this account, the first task is to set down some topics which ther author has found to be basic to entry-level MST TMA questions.

**MATHS TOPICS COVERED**

(1) Substituting values in formulae

(2) Rearranging simple formulae

(3) Solving linear equations

(4) (S.I.) units of measurement

(5) Powers of 10

(6) Scientific notation

(7) Decimal places and significant figures


A few general remarks about topic coverage in SGSG will be briefly considered next, followed by a discussion of each of the above topics in turn.

*SGSG* COVERAGE OF TOPICS
More basic concepts such as 'the four rules' of arithmetic and 'decimals', are included in SGSG.

However, the important topic of 'transformations',

is not covered. The former is a major topic in some MST courses and is a significant omission from this point of view. In one entry-level OU course which the author studied, it was the subject of a whole chapter in the text covering one module.

Discussion and examples of the individual topics listed above now follows. Those whose treatment in *SGSG* leaves little to be desired- such as rearranging formulae and solving equations- are quickly dealt with, others are considered in more detail.

(1) **SUBSTITUTING VALUES IN FORMULAE**

The use of simple formulae to determine the value of some variable is explained (in *SGSG*) in terms of substituting values in place of words or symbols in a formula, and then applying the 'rule' of the formula to calculate the result.

One example is given below. Leaving aside units of measurement for the moment, the problem posed is to use the formula:

$$a=(v-u)/t$$

to find a, given the values of v, u and t. The procedure is very straightforward. Since the above values are 12.5, 0 and 30, substituting them in place of v, u and t gives:

$$a=(12.5-0)/30$$

as an initial step in the working.
Interestingly, in one publication (Gilmartin *et. al.* (2001), p. 33), the result of this calculation is written down with full calculator accuracy as 0.416666666, even though t is given only to 2 figure accuracy. In most instances, *SGSG* manages to avoid quoting intermediate results to 9 figure precision by choosing values which give results exact to 2 or 3 figures. For example, if the values 10, 5 and 20 had been used for v, u and t, then,

$$a=(10-5)/20$$

$$=5/20$$

$$=0.25$$

would have been obtained. In the real world, of course, using actual data, results as neat as this would be unlikely to be obtained.

Another device used is to 'hold over' the results of intermediate calculations, quoting just the answer to the required accuracy. Although this procedure leads to expressions which tend to become unwieldy where there more steps involved in a calculation, using the original values for v, u and t in the example gives:

$\quad a=(12.5-0)/30 =0.417$ (to 3 figure accuracy),
thus avoiding the need to write down the intermediate result '0.416666666'.

Of course, 0.416666666 could have been shortened

to 0.41 6', with a dot over the 6 to indicate that this digit recurs. More generally, a suitable number of figures can be quoted trailed by three dots ("ellipsis") to show that further digits were generated in a calculator's display:

$$0.416...$$

Though, unlike the above recurrence method, this does not provide information about what these omitted digits are, which may be of some importance when checking through the steps of a calculation.

In spite of these differences, implicit in any correct approach is the retention and use of *full figure calculator accuracy in all steps of working* to avoid 'rounding' errors, only rounding at the end to give the result of the calculation to the degree of accuracy appropriate to that of the least accurate datum of the data used in the calculation (e.g. to 3 figures, if, as was assumed in the above calculation, them value for v is the least accurate datum).

(2) **REARRANGING FORMULAE**

The main point here as stressed by *SGSG* is the need to 'do the same to both sides of the equation'.

# Writing OU MST Assignments

## (3) SOLVING EQUATIONS

*SGSG* also gives several worked examples using 'word formulae', which may be prevalent in MST modules.

## (4) UNITS OF MEASUREMENT

I found the following units were used in the entry-level MST assignments in the course modules and I studied:

| | |
|---|---|
| Money | £ |
| Length | m |
| Area | $m^2$ |
| Angle | ° |
| Time | s |
| Frequency | Hz |
| Force | N |
| Pressure | $N/m^2$ |
| Volume | $cm^3$ |
| Density | $g/cm^3$. |

*SGSG* contains no reference to the fundamental unit of force, the Newton (N), in its index, which will feature in many MST assignment problems in mechanics. Neither is density mentioned, even though both these units featured in the assignments questions in the elementary MST OU courses the author studied.

One interesting difference in the approach to units in other publications and *SGSG* on the one hand and the modules studied by the author on the other, is in how they are used in calculations. *SGSG* and some other publications tend not to write in the units together with values in every step of the calculation, whereas modules may well do so. Returning to the above example, the writing of

$$a=(12.5-0)/30$$

rather than

$$a=(12.5 \text{ m/s} - 0 \text{ m/s})/(30 \text{ s})$$

illustrates this difference in approach. The authors of *SGSG* on p. 368 say that, whereas the units of measurement are left out of the intermediate steps of calculation 'out of habit', their inclusion in some MST courses is a *requirement* (which again was the case with some of the modules the author studied). An advantage of writing in the units throughout is that it encourages consistency in the use of units in the steps of working and those appearing in the final result, reducing the possibility of errors made on this account.

A disadvantage is that this practice adds

# Writing OU MST Assignments

complexity, as can be seen from the above example of one line alone, and the increased clutter can cause errors to be made. There seems to be a mathematics/science and technology divide on this issue.

## (5) POWERS OF 10

An understanding of this topic is basic to a grasp of subunits and scientific notation (topic (6) below). One approach to this topic is under the general head of 'Indices'; positive, negative and zero powers all explained in terms of a general base 'a', where a stands for any natural number.

This approach does offer a more complete understanding of powers of 10, which may be all that is required for some entry-level MST modules. Thus, using the rules for the multiplication and division of numbers raised to a power, described in the associated *Student Toolkit* booklet, negative powers may be understood as reciprocals of positive powers (e.g. $10^{-3}=1/10^3$**, since by the division rule $1/10^3=10^{(0-3)}=10^{-3}$). Similarly, using the division and multiplication rules, zero power may be understood as the result of dividing a number raised to a power by itself (e.g. $10^3/10^3=1=10^3 \times 10^{-3}=10^{(3-3)}=10^0$).

## (6) SCIENTIFIC NOTATION

Although scientific notation is covered in

SGSG, it could be made clearer that a number expressed in scientific notation has only one digit before the decimal point regardless of the size of the number represented. For this reason, 584 expressed in scientific notation is $5.84 \times 10^2$ and not $58.4 \times 10^1$. A wider range of examples which drives home this idea might also have been given.

Interestingly, *SGSG* does not take the opportunity to refer to the use of exponent notation, common in calculator and computer displays, which is a direct equivalent of scientific notation, easier to write down and more compact. For example, 56432 is $5.6432 \times 10^4$ in scientific notation and 5.6432E4 in exponent notation ('E' represents the "Exponent" or power of 10).

(6) **DECIMAL POINTS AND SIGNIFICANT FIGURES**

The difference between quoting a number to a given degree of accuracy in decimal places (d.p.) and significant figures (s.f.) is commonly misunderstood. Mistakes are often made in dealing with the presence of zeros at the beginning, at the end and in within the string of digits of a number.

Although *SGSG* otherwise adequately covers this topic, calculators do not necessarily solve the above source of difficulty. My calculator for example (a CASIO *fx-83MS*) can be set to display

# Writing OU MST Assignments 63

input and output to a given number of *decimal* places, but in order set the number of significant figures in displaying the result of a calculation, scientific notation must be used, thus *changing* the format of numbers keyed in (i.e., from decimal numbers to numbers expressed in scientific notation), whereas an output in the same format as input may be desired. For example,

$$3.7621 \times 4.3257$$

is output in normal display mode as 16.27371597. Switching to scientific mode and setting the number of (significant) figures to 3 produces the display:

$$1.63 \times 10^{01}$$

This does gives a result correct to the desired 3 significant figures but the result required is the *normal decimal form* the same as the operands (the numbers used in the calculation), *i.e.*, 16.3, not *scientific notation* as in the calculator's display, even though this is easily converted (though armed with a knowledge of Indices, the conversion from one to the other can be made readily for small values for the numbers of digits/s.f.s as in the above example, but if larger values have to be converted, errors can creep in in the process).

Those who are interested in elementary programming will find a program which converts numbers to a given degree of decimal place/significant figure accuracy in another book (Trigger (2014a)).

**CONCUSION**

Summing up, I found some publications recommended by the OU to be of some, albeit limited use as a supplementary source of tuition and examples in the maths topics which commonly feature in TMAs associated with elementary MST courses. For example, I found the explanation of scientific notation in the *Student Toolkit* series helped to clarify the exposition of this topic in *SGSG*. Supplementary materials, if they focus on the material which is pitched at an appropriate level, may be useful to students seeking extra help and advice with the maths topics in such modules which will likely feature as subject matter in associated TMA questions.

The omission of transformations as a topic in these publications and *SGSG*, was also viewed as a significant drawback.

---
\* The Sciences Good Study Guide

\*\* The symbol ^ is used to denote that the following digit is a power. E.g. 10^3 is exactly the same as $10^3$ in standard notation. It is frequently used as a substitute for standard notation where limitations in electronic media require it.

.

Sometimes, a TMA mathematics-type question may ask the student to derive a formula to fit a set of given data, which may often be determined by inspection of the data for a pattern. In a more advanced course, the question may then go on ask for a mathematical proof that the formula derived is true in general (i.e. in *all* cases). These are dealt with by example in the next chapter.

# CHAPTER 4: MATHEMATICAL DERIVATIONS AND PROOFS

## DERIVATION OF A FORMULA FOR THE NUMBER OF BEDS IN A RECTANGULAR GARDEN

The middle figure on p. 111 of the *Sciences Good Study Guide* is drawn below:

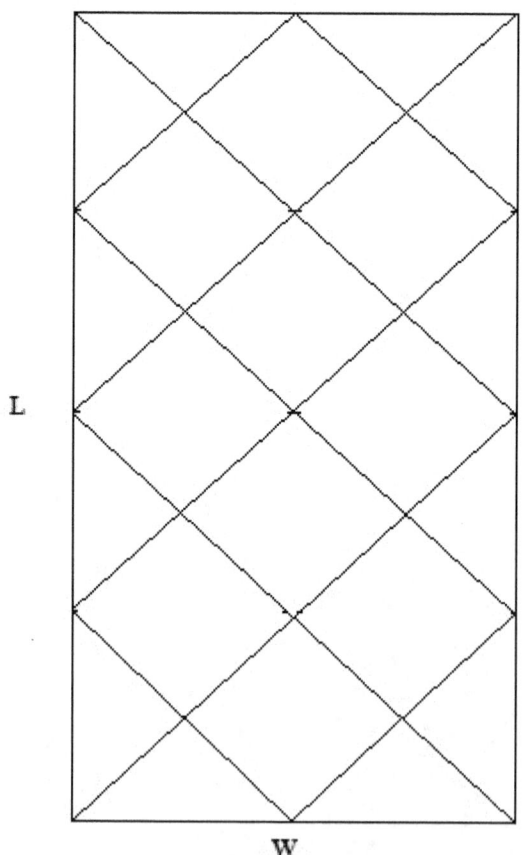

Figure showing a rectangular garden with flower beds, L beds long, W beds wide.

Suppose there is a single row and a single column of beds. From p. 32, Exploring Pattern, the number of beds is 3L+1=4, since L=1.

Suppose now that the area is extended downwards and to the right.

If the width of the area is W, this will involve adding W-1 rows. Each of these rows will add 3L+1 beds, minus those L beds which are simply replacements of triangular beds with square ones, giving 2L+1 added beds.

Hence, we have:

N= 3L+1 +(W-1)x(2L+1)

Simplifying (expanding the brackets by multiplying the first bracket (2L+1) by the W then by the -1 of the second bracket (W-1)), gives:

N= 3L+1 +2LW +W -2L -1.

Collecting like terms in the variables L, LW and the constants 1 and -1,

N= 3L-2L +2LW +W -1+1.

So **N=L+W +2LW**, which is the number of flower beds in a rectangular garden with L rows and W columns of beds.

PROOF OF THE EXPRESSION N=L+W +2LW

(The maths involved in this proof is Second Level Mathematics. It is included for the sake of completeness. Readers who are more advanced in their mathematical studies may find this of interest as an example of a proof by mathematical induction, but beginning students should not be surprised or concerned if at their present stage of study at the OU they are unable to follow the argument).

By reference to the figure and the derivation argument above, first note that adding one row increases the number of beds by 2L+1, and adding one column increases the number of beds by 2W+1.

Let p(L,W) be the proposition that N(L,W)= L +W +2LW, L,W=1,2,3,...

Let p(L,1) be the proposition that N(L,1)= L +1 +2xLx1 =3L+1.

N(1,1)= 4 by inspection. p(1,1) is that N(1,1)=1 +1 +2x1x1 =4. Hence p(1,1) is true.

Suppose p(K) is true, i.e., that

$N(K,1) = K + 1 + 2 \times K \times 1 = 3K+1$ is true.

Then the proposition $p(K+1, 1)$ that
$N(K+1, 1) = K+1 + 1 + 2 \times (k+1) \times 1 = 3k+4$
is true for $K = 1, 2, 3, \ldots$

But $N(1,1)$ is true, so $p(L,1)$ is true for $L = 1, 2, 3, \ldots$ by mathematical induction.

From above, $(K,1)$ is true for $K = 1, 2, 3, \ldots$
Suppose $p(K,M)$ is true, i.e., that $N(K,M) = K+M+2KM$ is true.

The proposition that
$$p(K,M+1) = N(K,M+1) = K + M+1 + 2K(M+1)$$
$$= K + M+1 + 2KM + 2K$$
$$= K + 2KM + M + 2K + 1$$
$$= K + M+1 + 2K(M+1)$$

is true for $M = 1, 2, 3, \ldots$ But $p(K,1)$ is true for $K = 1, 2, 3, \ldots$ Hence $p(L,W)$ is true for $L, W = 1, 2, 3, \ldots$

Since $p(K,M)$ is that $N(K,M) = K+M+2KM$,
**$N(L,W) = L+W+2LW$ is true for $L, W = 1, 2, 3, \ldots$**

.

Having considered some fundamental mathematical aspects of MST TMA *subject matter* conceptual content in Chapters 3 and 4, in the following chapters we shall move on to consider the main elements of their *pictorial* content- graphs, tables and mathematical charts in Chapter 5, and 'biological' charts and diagrams in Chapter 6.

# CHAPTER 5: CONSTRUCTING EFFECTIVE MST GRAPHS AND TABLES

**INTRODUCTION**

This chapter is concerned with Charts, Graphs and Tables, which commonly feature in MST TMAs.

The advice given covers the presentation of data in the form of charts and tables as well as graphs and may be used as a source of material for all or any of these. But the principal focus here is on graphs.

Both the *construction* and *interpretation* of charts, graphs and tables are considered, since both may be involved in an MST TMA question. In particular, we will be concerned with the strengths and weaknesses of examples of such diagrams in presenting data.

As forms of visual representation of data, the charts, graphs and tables discussed here, a number of general points which emerge from the discussion, are relevant to the construction of effective charts, graphs and tables graphs for OU MST assignments. In the following paragraphs, these points will be considered in relation to the steps of construction. Then, some brief remarks will be addressed to making proper use of such figures in the text accompanying it, the main reference is to graphs, but charts and tables will also be referred to.

## (1) THE DATA

*THE SUITABILITY OF THE DATA*

Graphs are most effective in showing the **relationship** between a **dependent variable** and an **independent variable**. Distance-time graphs are examples. In this case, distance is the dependent variable and time the independent variable:

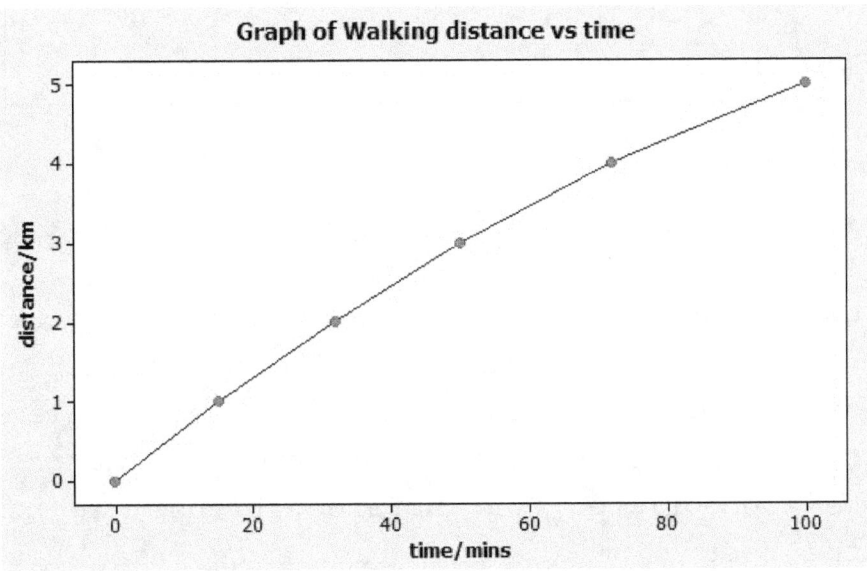

*Source: Observations based on the author's personal experience*

In another example the data are the size of the 'over 55 'workforce collected over a period of 25 years. In this case, the use of a graph is appropriate to show a decline in the 55+ workforce over time:

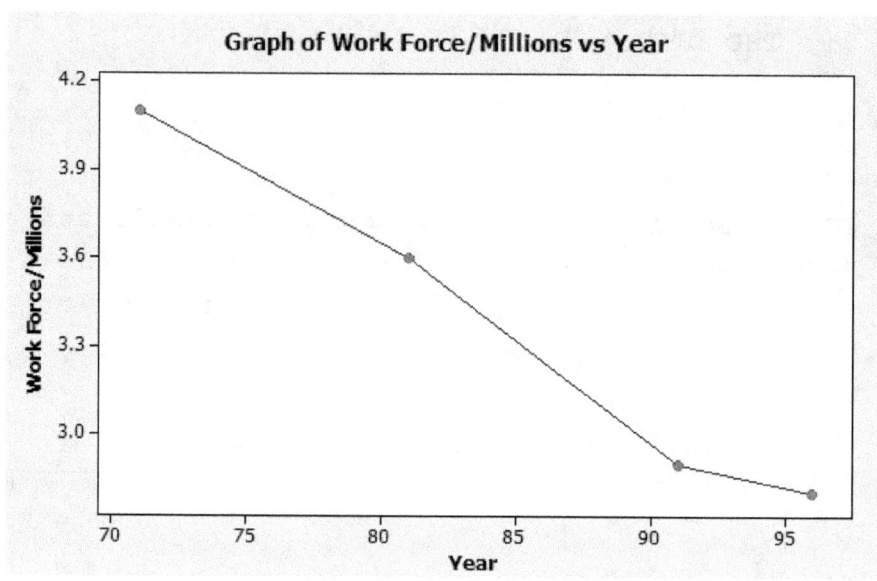

Source: Student Toolkit 3, p. 23

Where there is no clear relationship between variables, or where more than two variables are to be represented on the same diagram, graphical representation of the data may not be appropriate. For instance, graphs are not the best form of presentation of data which is in the form of proportions of a whole. An example is where the data are sizes of subpopulations as proportions of the whole population. Here, the use of a 'pie' chart to display the data is more suitable than a graph:

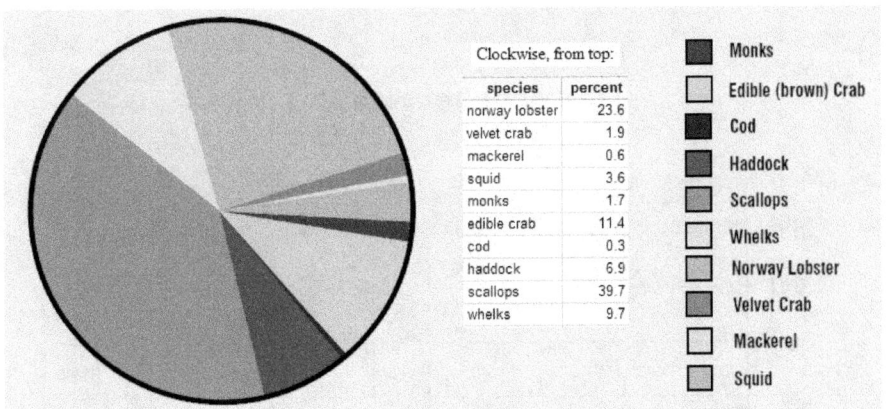

**SPECIES OF FISH CAUGHT IN AN AREA OF THE NORTH SEA**
*Source:* FRS (2002) *in* Talisman Energy (2005)

In this case, the 'population' is the total catch of fish in a particular region of the North Sea; the 'subpopulations' are the various fish species caught. [N.B. Because of rounding, the individual percentages may not add up exactly to 100% (in this case 99.4%)].

Where it is necessary to depict the relationship between a dependent variable and several independent variables, other forms of presentation may again be more appropriate than graphs. The depiction of data consisting of the distance travelled by the average person (the dependent variable) on various modes of transport (independent variables), is one example where bar charts are used in preference to graphs:

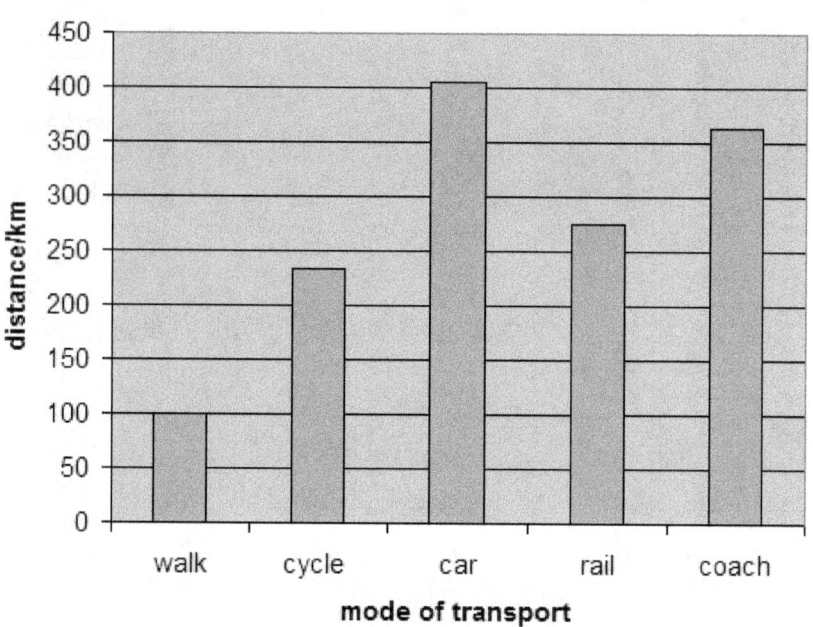

*Source:* Fictitious Data

One or more different sets of data may be represented on the same graph, perhaps for comparison purposes, provided the same scale can be used. However, the use of different scales for different data sets displayed on the same graph tends distort any relationship between them and to add confusion. Hence, conclusions drawn from such a graph based on visual comparison may be invalid. In the graph below, for example, the manufacturer of washing powder brand y wants to convince us that though a competitor's brand x was selling better up to 1995, after that time sales of brand y began to outstrip sales of brand x, and this *looks* from the graph to be so (the blue line representing sales of brand y rises well above that of the green line for brand x after 1995. But in fact it did not, as scrutiny of the scale belonging to brand y which is on the right shows; that for brand x is on the left. After 1995, brand y was still only selling under 4,000 items (referred to right hand scale), whereas brand x sold over 15,000 (left hand scale):

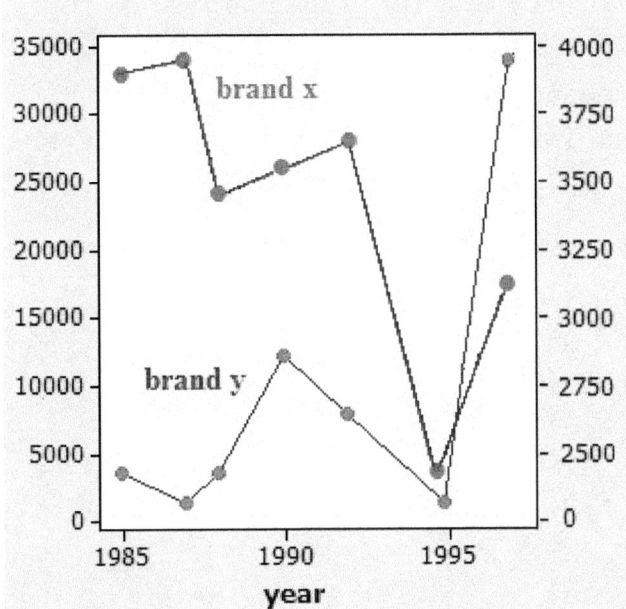

*Source:* Fictitious data

- It emerges from a consideration of inaccurate and misleading diagrams that actual data values rather than converted data values (taking percentages of data and then plotting the percentages is an example of the latter) tends to produce the most effective graphs. However, data values which have been transformed in some way, perhaps to produce a straight line relationship, may be acceptable for this purpose. For example, in the following graph, the power taken from the mains by 100-watt light bulb was measured as the voltage of the mains varied from 230 volt. When power vs voltage variation was plotted, the result, shown by the black line, was not straight. However, when the voltage variation was transformed by squaring and then plotted against power, a better straight line, superimposed on the same graph for illustration purposes [the values on the 'y' axis should also be squared] shown by the red line, was obtained:

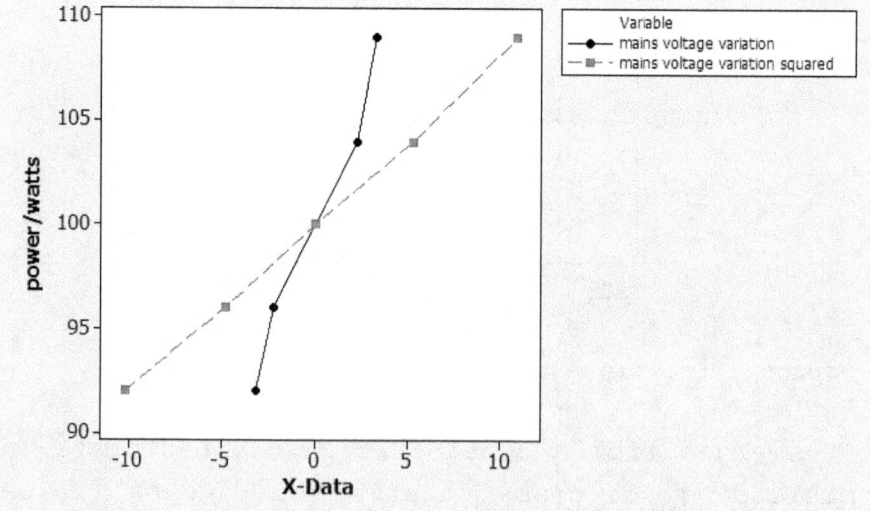

*Source:* measurements undertaken by the author

But where *data are converted to percentages or proportions,* when plotted on a graph, they can give a false impression of the strength of a trend in the data, especially if the y axis does not start at zero:

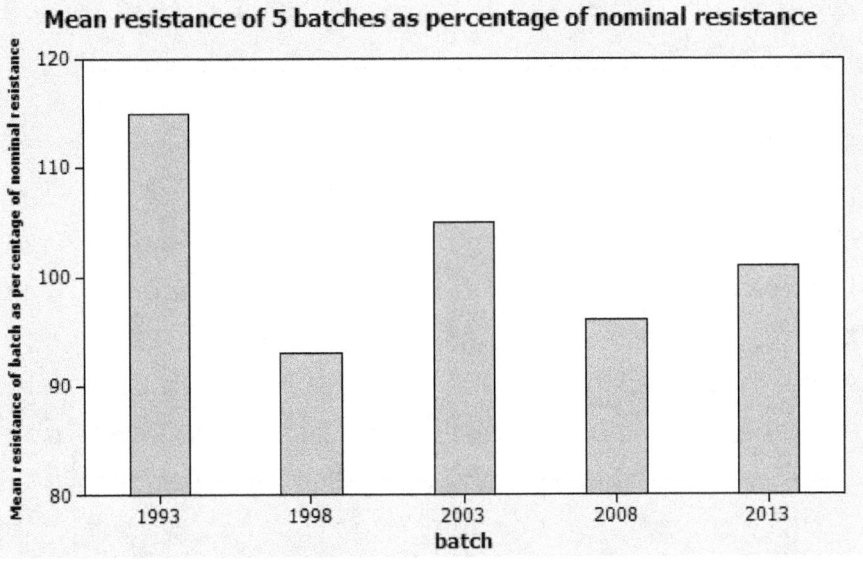

*Source:* author's measurements

When the data are replotted on a second graph (below):

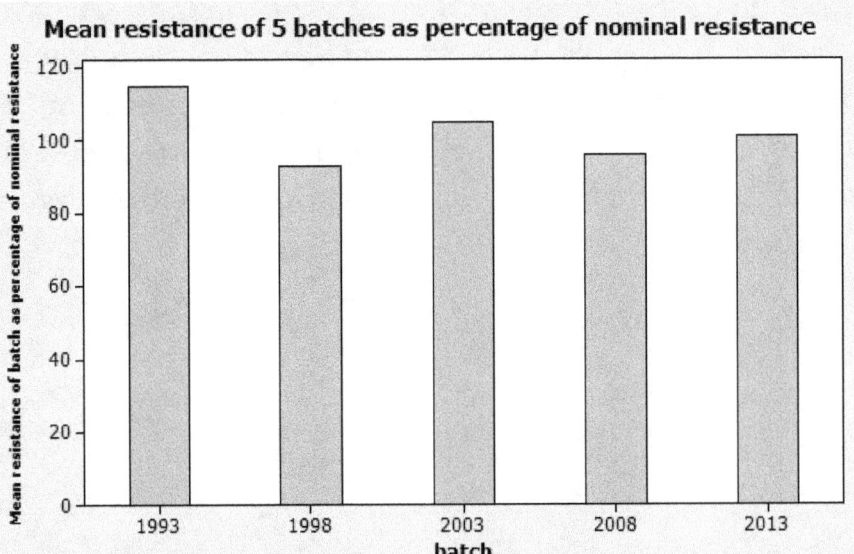

with a starting value of zero on the y-axis, and compared with the first graph, it is evident how the first graph visually exaggerates the trend of the mean resistance of later batches towards the nominal (i.e. the marked) value.

Furthermore, the *use of percentages or proportions* tends to *mask the amount of data* on which such interpretations are based, and therefore the extent of the *evidence for* any trend identified. In fact, from 2003, batch size was increased from 50 to 200, which means that the first two batches were more likely to give a mean value differing more widely from the nominal resistance value than the three later ones.

The *use of differences between data values* to plot a graph also tends to exaggerate apparent trends in the data. The effect is similar to that produced by using non-zero starting values on the axes, as exemplified below:

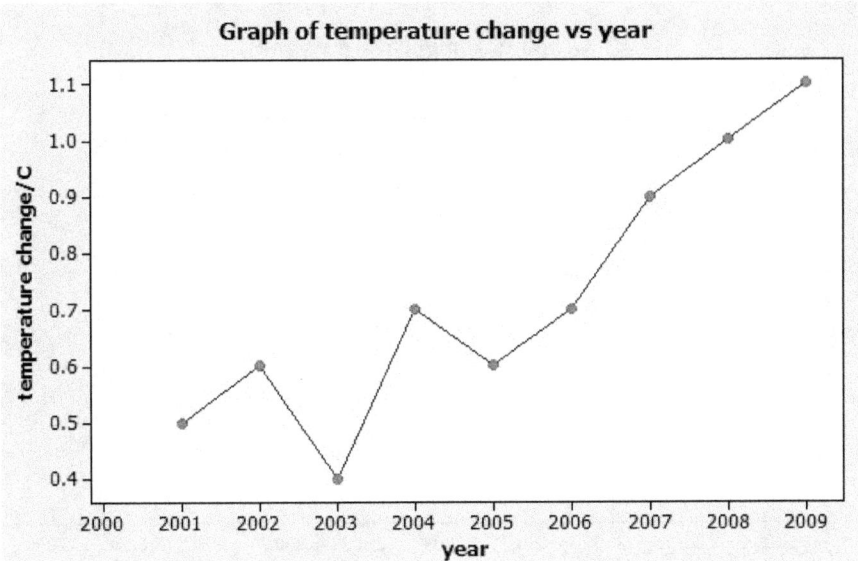

Source: Trigger (2013)

The above graph was used as evidence of 'global warming', apparently showing a strong trend in elevating temperatures relative to the year 2000, especially since 2006. However, the trend is seen actually to be much less [though an upward trend is still evident] when actual *recorded temperatures* instead of *temperature differences* (= 'temperature change' on the graph) were used in plotting the graph:

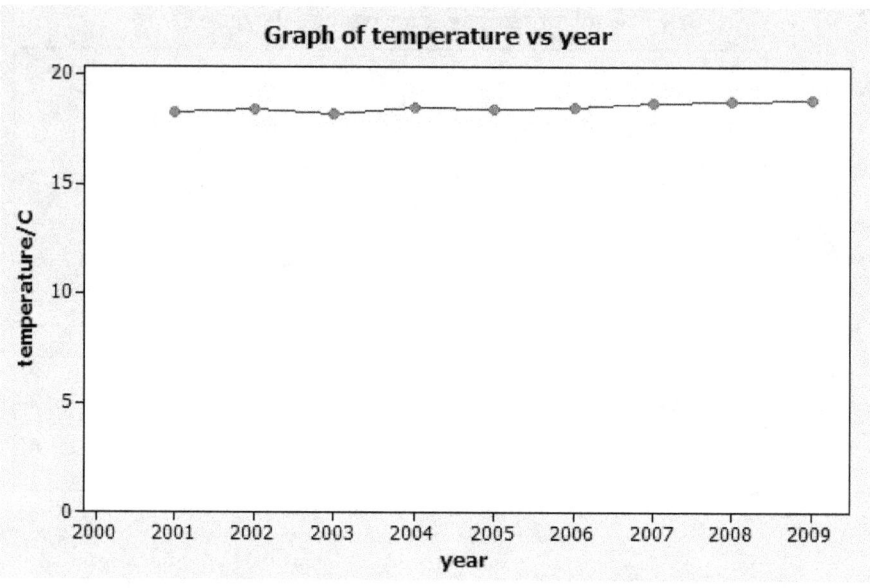

Another procedure and example of bad practice which tends to exaggerate an apparent trend in the data is the use of unequal intervals between values on the x-axis. In the following example, notice that the interval between the successive values of 1985 and 1992 marked on the x-axis is 7 years, but between those of 1997 and 2010 it is 13 years, yet they are represented by equal length intervals on the x axis, which exaggerates the appearance of the trend on the graph:

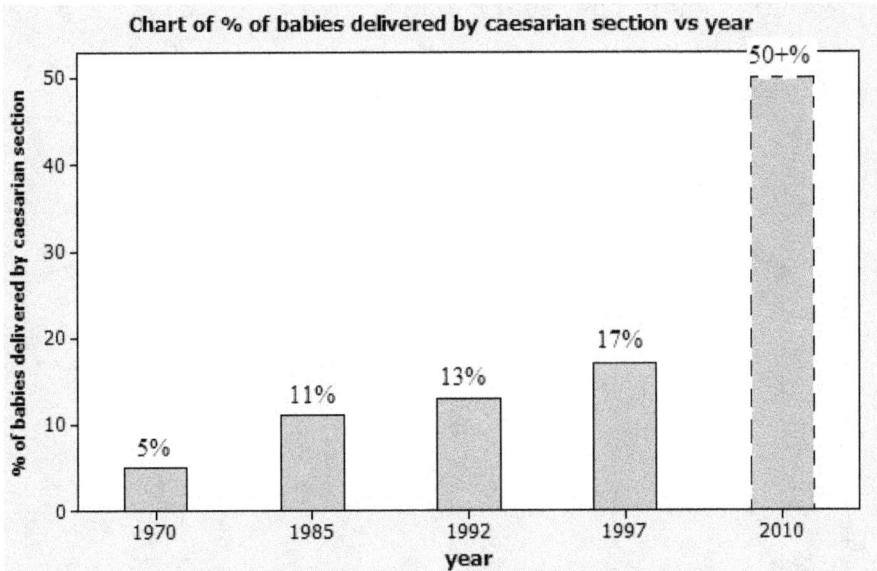

Source: *Student Toolkit* 3, pp. 27-28.

The percentage figure for 2010 depicted with a dotted bar added to the chart for the data itself is an *extrapolation* from that data for 1970-1997. Because of the wide interval in time between 1997 and 2010, and the fact that the percentage of babies delivered by caesarian section has been increasing since 1970, the projection for 2010 looks feasible at first glance. But looking more closely at the figures given, one feels that it may be exaggerated. The comment of the author in which the bar chart appeared said this:

> "One specialist I spoke to reckons that the rate could rise to over 50% within the next 20 years. A quick look at the graph suggests he may be right."

However, when the chart was redrawn as a line graph with an evenly spaced scale on the x-axis, the following figure was obtained:

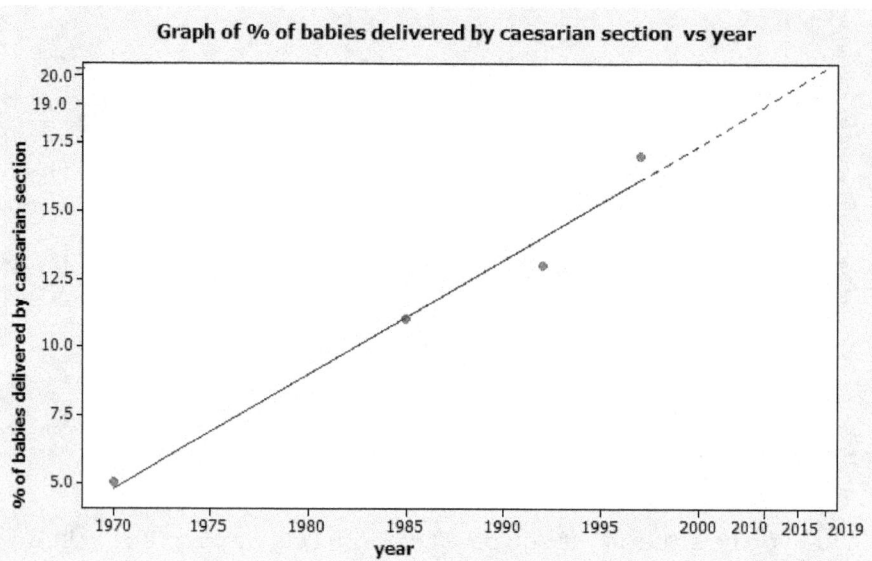

The trend line (a line of 'best fit' to the data points known as the linear "regression line") shows that the expected rise in the percentage of caesarian births in 2010 is 19.0%. Now the article from which the above quote is taken is dated January, 1999. So the prediction of a rise of 50+% in the above quote refers to 2019. However, the trend line points to an expected rise of 20.3%, *not* 50% as shown on the bar chart.

**THE DATA SOURCE**

It is important to provide information, written on the graph, about *where the data used to plot it has come from*. This should be sufficiently detailed *to enable the reader to check the accuracy of the graph*, and to assess the validity and reliability of any interpretations made from it.

The graph itself should *indicate the source* briefly, but appropriately detailed information should be given in the accompanying text or list of references. However, in the case of graphs plotted in answer to set questions, the necessary information will be included in the question, and so the reader needs only to be referred to this.

(2) **TITLING AND LABELLING**

The main purpose of the title is to state what the figure shows. Though it is common in the newspaper and magazine articles featuring in the *TOOLKIT* series to use a 'catch phrase', any additional heading seems unnecessary in academic work, like MST TMAs. A simple, but sufficiently explanatory title is best. This should be clear, concise and prominently displayed at the top of or underneath the graph. The title might be taken from the source heading, such as a table from which the data were obtained, perhaps also referencing the data source at the same time. Otherwise, the source may be referenced directly under the title, or immediately below the graph.

In the case of graphs and bar charts, the description of the values on the axes is no less important than its title. A single word, phrase or a short sentence is all that is necessary and desirable in describing the variable represented on each axis, *together with its unit of measurement* where appropriate.

The numbers written against graduation marks along the axes, which define the scales, should be *centred on each mark*. These numbers need to be appropriately spaced. If this is not the case, it may be difficult to determine which number refers to which mark. An example is where numbers representing values along the horizontal axis overlap more than one graduation mark, making the graph difficult to read properly. This difficulty can be avoided by the use of such devices as choosing larger scales for the axes and writing the values on the axes more concisely or against alternate marks. As an example of the second device, the marks on an axis running from 1 000 to 10 000 could, if necessary, be labelled from 1 to 10, and the factor 'x 1 000' added to the axis description. Slanting or vertical scale values may be used though they may reduce the space available on the page for the graph itself, and are more difficult to read.

Whatever title and labels are used, they should be chosen and displayed on the graph so as to *minimize clutter and maximize readability, a principle which is* **applicable to the whole assignment.**

## (3) SCALING THE AXES

Scaling refers to the size of the interval used to graduate an axis, represented by some unit of length on the graph paper. (This unit is 1 cm if centimetre graph paper is used). *SGSG* urges making full use of available space in scaling the axes of a graph. This facilitates the use intervals which make data points easy to plot, values easy to read from the graph and relationships in the data easier to quantify.

Most graphs use scale intervals of 1 or multiples of 10. With cm/mm graph paper this enables values to be easily located along the axis. Intervals in multiples of 2 or 5 are less convenient in this respect, but may be necessary to make full use of available space for clarity. It is apparent that the use of more awkward scale intervals makes accurate plotting and reading of values from the graph more difficult and more prone to error.

As in the case of bar charts, pitfalls occur if graphs are constructed with axes which do not start at zero. If for example, values of a dependent variable range from 80 to 120 and the scale on the corresponding (vertical) axis starts at 80 instead of 0, the effect is to exaggerate the changes in the dependent variable which occur with changes in the independent variable. We have seen that such differences are much reduced when the same data are plotted on a graph with the vertical axis starting at zero.

Another problem arises where data values are confined to a range distant from zero. The use of axes starting at zero in these situations would waste graph space and tend to reduce accuracy when plotting the data points. The accuracy of any line drawn through them would also tend to be reduced by the resulting 'crowding' together of data points. However, a scale extending down to zero may not be appropriate to the possible range of the data depicted (an example is the 'hominid brain size' graph on p. 99). For these reasons, it may be necessary to start one or both axes with appropriate non-zero values, or install an axis 'break' (see the y scale on p. 99).

(5) **PLOTTING A GRAPH**

DISCRETE/CONTINUOUS DATA

Both **discrete data** and **continuous data** can be effectively displayed using graphs. Discrete data refer to items which can be *counted*. This includes items which fall into one of several categories (categorical discrete data). Some examples are: calendar years; items grouped together and the groups numbered, and number of individuals or objects. The 'modes of transport' referred to earlier on p. 74 is an example of categorical discrete data.

When discrete data are plotted on a graph, the points should *not* be joined with a line. This is because intermediate values such as

'2.5' on an axis representing, say, 'number of individuals', are clearly meaningless. Another good example where data points should not be joined with a line, is a graph depicting the 3 times table.

Certain types of *continuous* data may be used to plot a graph where the points can be joined together with a line. This is valid where it is clear that the relationship between the graphed variables applies in the intervals between data points. For instance, the relationship between the area of a square and its side length$^2$ (*i.e.* length x length) is known to apply for all values of side length, and so the points on a graph of area vs side length$^2$ may in this case be joined with a line.

Where such a relationship is not clear, it may not be appropriate to join data points in this way. An example is a line graph connecting data points representing the values of a dependent variable for each *year* of interest (the independent variable). Where this is subject to *seasonal* variations, connecting yearly data with a line may render dubious the representation of the behaviour of the dependent variable BETWEEN years.

**EXTRAPOLATION FROM A GRAPH**

An example of the pitfalls of extrapolating from a bar chart was discussed above. But now consider an example where a graph is extended into a region for which there is *no* data. As in graph (d) on p. 78 of *SGSG*, this extension takes the form of a dotted line, emphasizing some *preferred interpretation* of the data. This device suggests that extending a line graph beyond the range of the data collected may not be appropriate. Instead, it may be desirable to explain, in the accompanying text, how the behaviour of the graph in regions where there is no data might be predicted from trends identified in regions of the graph where there are data. An example is in the answer to the following MU120 TMA question:

> "The number of children in a small town owning the latest 'must have' toy is shown in the table below:

| *TIME* | *T/weeks* | 1 | 2 | 3 | 4 | 5 | 6 | 7 |
|---|---|---|---|---|---|---|---|---|
| Number of children owning the toy | N | 3 | 5 | 8 | 13 | 21 | 33 | 55 |

> [the next part of the questions asks students to plot a graph for this data and to use the graphics calculator *Texas TI-83 Plus* which goes with this course to determine a formula for the graph which turns out to be $N = 1.88 \times 1.62^T$]

> "In which week does the model predict that the number of children owning the

toy would first exceed 500?

Comment on the reliability of this prediction."

ANSWER

The graph plot of the above data is shown below, followed by a comment about the goodness of fit of an exponential curve to the data. This is followed by a graph with an extended x-axis to allow of *prediction beyond the given data range*, and an explanation as to why prediction beyond 25% (as a rule of thumb) of the data range might in this case lead to an unreliable estimate:

P. Trigger    T3730911    MU 120    TMA3

Q.4(i)

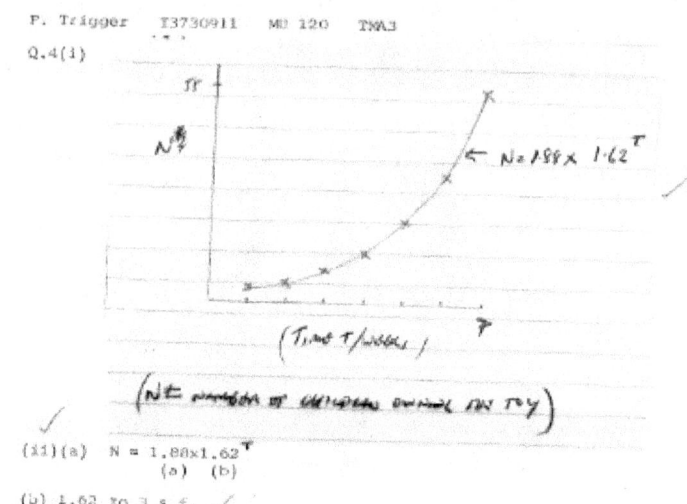

(ii)(a)  $N = 1.88 \times 1.62^T$
         (a)  (b)

(b) 1.62 to 3 s.f.

(c) After the first week, there were 3 children owning the toy; subsequently, the number of children owning the toy is increasing by a factor of about 1.62 from one week to the next.

(d) The correlation coefficient in this case is a measure of how well the regression line fits the data. Here, $r = 0.9999$ to 4 s.f. . As this value is very close to 1, the fit is very good (now proper).

(1) I think as the craze of the toy catches on, more and more children are likely to own it as time goes on. So, at least to start with (in the early weeks), I would expect a growing upward trend in toy ownership which is more akin to a power curve than a straight line relationship between N and T.

(2) The scattergraph clearly shows that any straight line drawn through the plot would be more distant, overall, from the plotted points than would an exponential curve. I.e., it is clear that an exponential curve provides a better fit than a straight line.

(iii)(a) Since $N = 1.88 \times 1.62^T$, substituting $T = 9$ gives an estimate of how many children own the toy after 9 weeks (using the regression function).

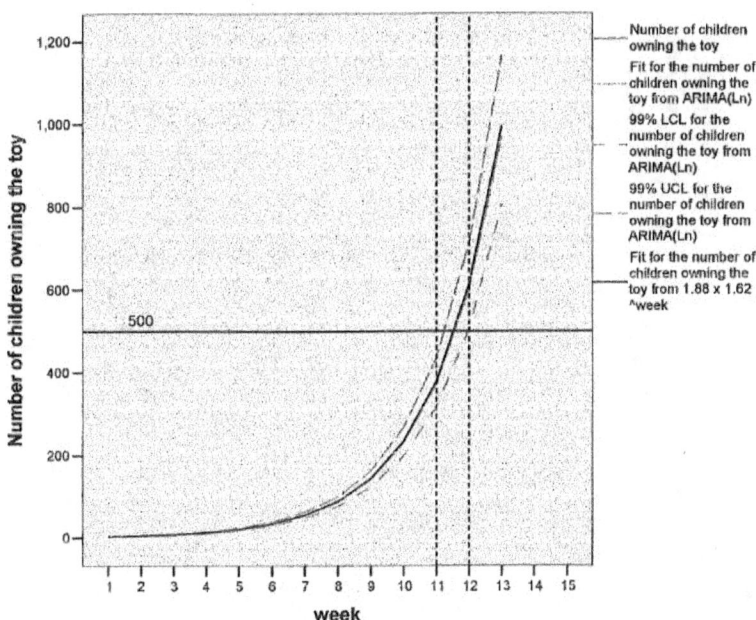

**Prediction of the week in which toy ownership will reach 500**

A software package (SPSS) was used to plot the exponential curve fitted to the 'toy' data (shown above), extended beyond the data range (which is 1-7 weeks), in order to predict the week when ownership of the toy will first exceed 500. The SPSS prediction is that this will be between the 11th and 12th weeks with 99% certainty (between the dotted verticals on the graph).

> P. Trigger T3730911    MU120    TMA3
>
> Q.4.(iii)(a) (cont)    Hence N= 1.88 x 1.62$^9$, which is approximately equal to 144.
>
> (b) I switched off the plots then plotted the full equation for the exponential curve in 'graph'. Then I used 'trace'. I found that for X= 11.489..., Y= 472.92... and for X= 11.702..., Y= 523.87... .
>
> Hence the model predicts that toy ownership rises above 500 in the 12th week (that is, after the end of the 11th week but before the end of week 12).
>
> If the modelled relationship and the variation in the number of children owning the toy continue to hold, the near-perfect (r ≥.9999) fit of the model predicts that toy ownership first exceeds 500 in week 12 (which is also not very far outside the data range on which the model is based) with very high reliability.
>
> In reality, the model may not continue to hold in the time interval of interest (weeks 8 - 12). Ultimately, the rate of increase in toy ownership must decrease (for example, when many of the eligible children already own the toy or when the craze dies down). In a small town, this is likely to happen sooner rather than later and if it occurs during the time interval of interest, the above model would predict too low a value for the number of the week when toy ownership would exceed 500 for the first time.*(f t m)* On the other hand, if for some reason toy ownership 'explodes', the above model would predict too high a value. In either scenario, then, the prediction would be unreliable.
>
> *or state that extrapolation is beyond 25% of the data range*

•

So, even where the data appear to closely follow a particular mathematical form, one should not go too far outside the data range in making predictions from it. In the above example, if the rule of thumb suggested by the tutor is adhered to, this gives:

25% x 7+7= 8.75

and one should not attempt to predict toy ownership beyond the 9th week.

Writing OU MST Assignments             93

A detailed example of an answer to a MST assignment question involving the construction and interpretation of graphs is described below.

## A TMA 'GRAPH' Question Y155

QUESTION:

"A space probe has discovered a lump of alien material...

(a) Using the data [given] plot a graph [GRAPH 1] of force vs extension [vertical axis scale is to be: 5cm to 10 N and 2 cm to 0.5mm on the horizontal axis]. Calculate the slope of the line.

(b) By comparing your graph [GRAPH 2: with lines drawn on another graph for each of the other materials using data given], decide whether the alien material is most like a ceramic, polymer or metal (Maximum **100 words**)."

ANSWER:

"(a) Graph 1 is shown below. Its slope is given by:

change in force/change in extension

= 40/2.67 N/mm   (see red line on graph)

= 14.98... N/mm.

So the slope of the graph is 15.0 N/mm.

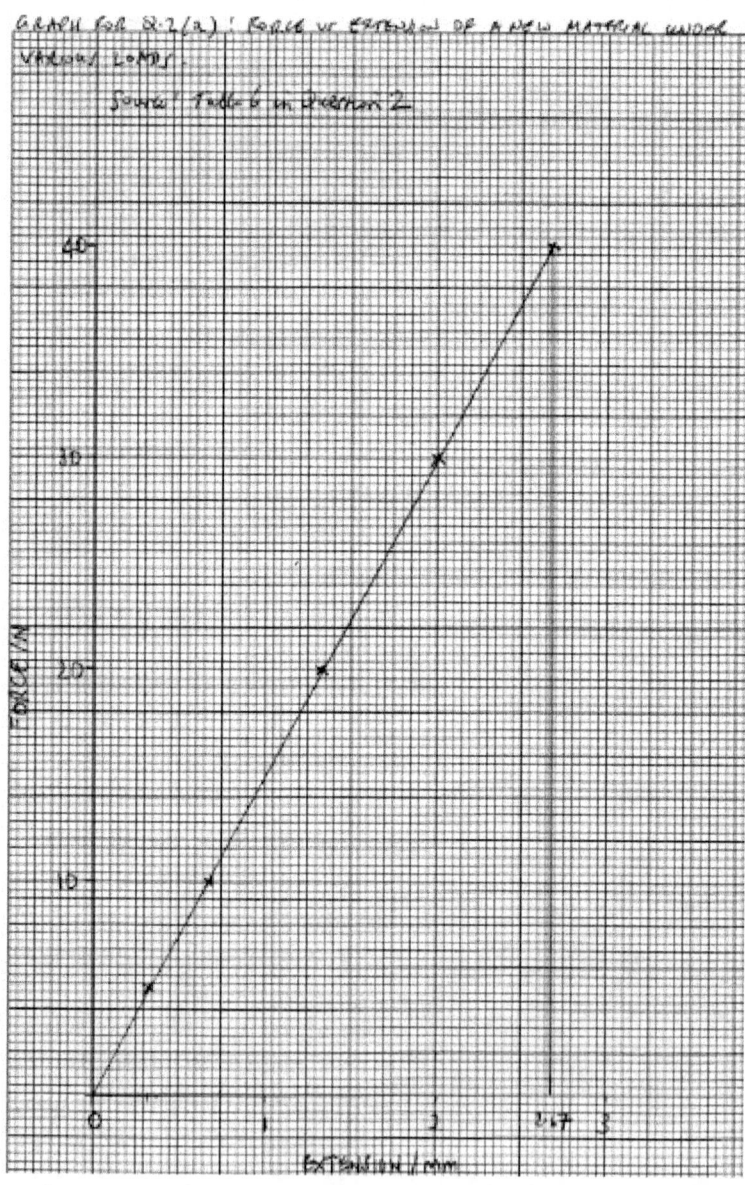

GRAPH 1

(b) Graph 2 shows that a polymer (polyethylene) is counted out because it cannot be strained beyond 1% of its original length, giving its line a much shallower slope than that for the alien material, whereas the alien material can be strained by 4% [Table 3, Y155 *Assignment Booklet)* at least:

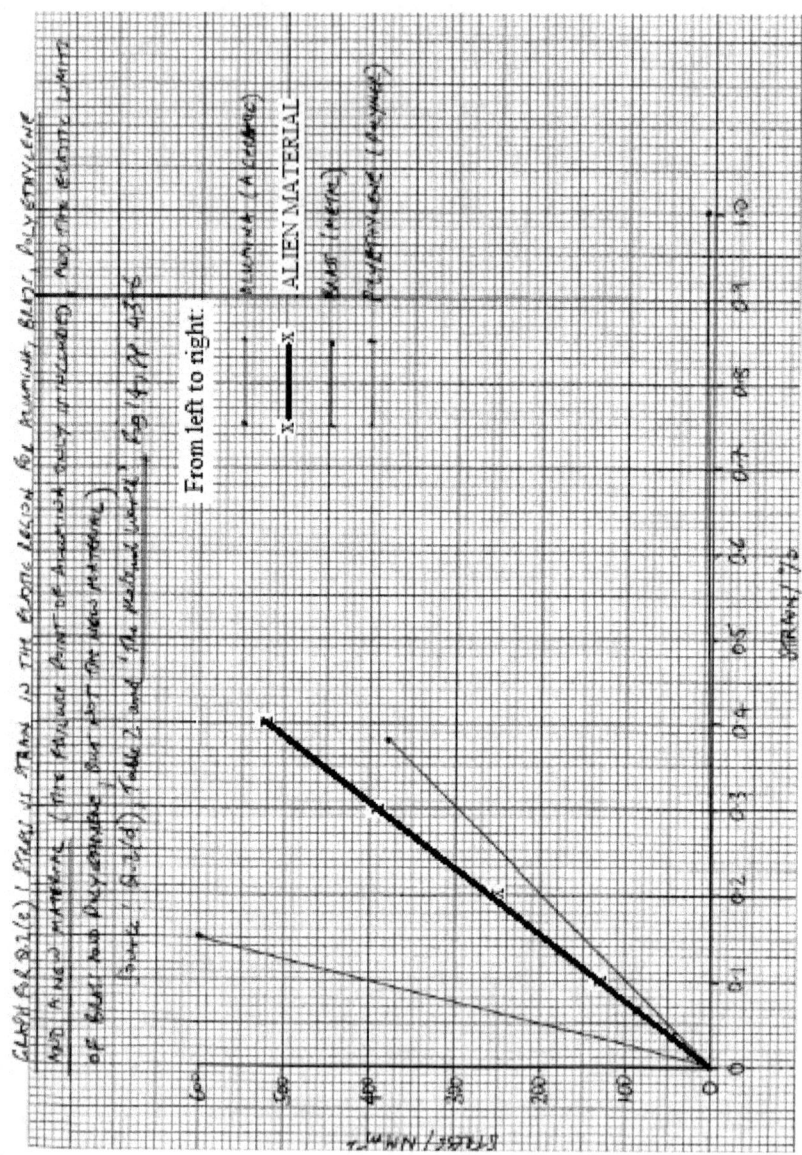

GRAPH 2

[Note that to make best use of the space available for the reasons already discussed, it may be necessary to draw the y-axis along the short side of the graph paper].

The slope of the stress vs strain graph for a ceramic (alumina) is much steeper than that for the alien material. Although the elastic limit for a metal (brass) is less than that for the alien material, the slope of its line is closest of the three. So, I think the alien material is a metal."

•

## COMMENTS

Features to Note

Graph 1 is headed with the title and data source. The independent and dependent variables are labelled alongside the axes together with their units of measurement. The FORCE and EXTENSION values used in the calculation of its slope are made clear on the graph itself, and of course, this fact is noted in the description of the calculation, so that the reader is directed to it (otherwise it might go unnoticed).

In Graph 2, The graph is made as large as possible given that all the data points for all four materials must be included and that convenient scale intervals (making points easier and more accurate to plot) should result. This makes for greater accuracy, though the full length and width of the graph paper *could* have been used, up to 26 cm for the STRAIN scale for example. But this would have resulted in more awkward scale intervals of 0.038...%/cm (instead of 0.5%/cm).

A *Title* for the graph is given, saying what it shows and its *source*. The 4 lines are distinguished in different colours and are in addition identified by a key ("legend"). *The data points used to construct the graph are clearly shown* on it, so that the reader can judge the basis for and accuracy of the line drawn.

**NON-LINEAR GRAPHS**

Of course, though the graphs spoken about here so far have been mostly straight line graphs, an assignment might require the drawing of a non-linear graph. An example, from a S292 assignment, is shown below, in connexion with the evolution of the human brain from hominoid ancestors:

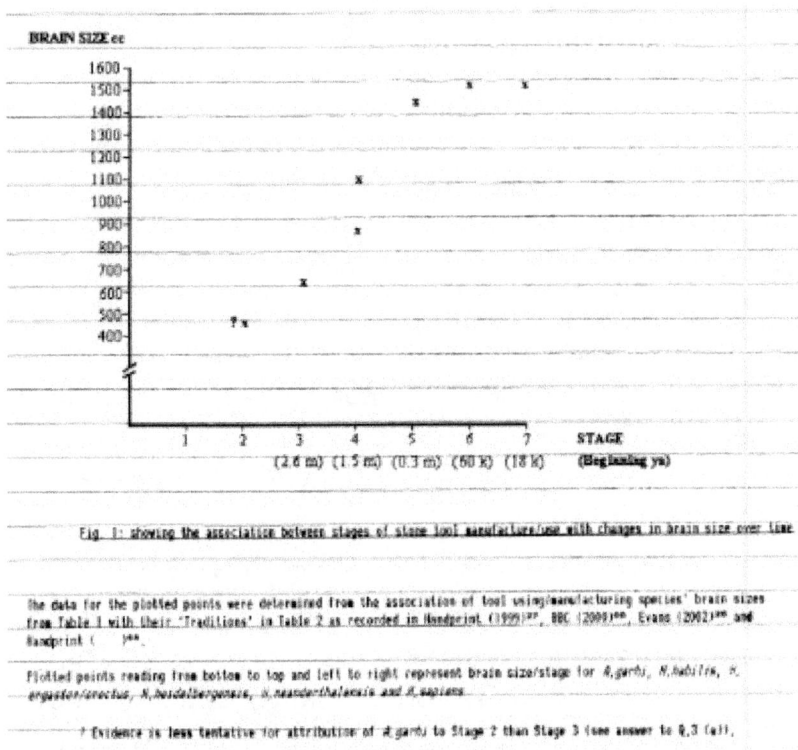

Fig. 1: showing the association between stages of stone tool manufacture/use with changes in brain size over time

The data for the plotted points were determined from the association of tool using/manufacturing species' brain sizes from Table 1 with their 'Traditions' in Table 2 as recorded in Handprint (1999)[aa], BBC (2000)[ee], Evans (2002)[ee] and Handprint ( )[aa].

Plotted points reading from bottom to top and left to right represent brain size/stage for *A.garhi*, *H.habilis*, *H. ergaster/erectus*, *H.heidelbergensis*, *H.neanderthalensis* and *H.sapiens*.

† Evidence is less tentative for attribution of *A.garhi* to Stage 2 than Stage 3 (see answer to Q.3 (a)).

MAKING USE OF A GRAPH IN THE ACCOMPANYING TEXT

Having discussed the construction of graphs, it remains to consider what use might be made of a graph in the text accompanying it. There are instances where the accompanying text fails to make proper use of the data presented. One example of bad practice is where the data presented at the beginning of an assignment answer are not referred to until the final paragraph of the answer. Another is where other data, *not* presented, are used in support of what the author is trying to show.

So, an important point emerging from these examples is that the accompanying text should make full use of the information shown in a graph, but it should not go beyond the data presented. The reason for this is that readers cannot check interpretations or conclusions based on absent data.

## TABLES

Tables are particularly useful for summarizing information. This may be numerical or textual, or a combination of both. However, when there is a lot of information to summarize, necessitating the use of more than a handful of columns and, to some extent, rows- if it is desirable to have the whole table on the same page (if not the table can be continued on further pages, though this can make intratable comparisons less direct and convenient), the resulting table can be unwieldy and difficult to read and draw conclusions from.

We shall see an example in an S292 assignment where the table included in the answer tends to swamp the reader with too much information, making the table rather 'indigestible'. Use of a minimum number of rows and columns results in the most effective table: 8 *columns* is probably about the limit. Simple column and row headings are best, which show the nature of the data and any units (or if the units are same throughout, clutter can be reduced by including them in the table's main heading).

To illustrate, firstly, an example of some of the pitfalls of constructing tables for MST assignments is discussed. This will then be followed by an example of well-constructed table.

The following shows an example of a poorly conceived and constructed table:

**THE CHANGING FACE OF STUDY AT THE OPEN UNIVERSITY**

|  | Now compared with 2012 | | | Projection for 2016 compared with now | | |
|---|---|---|---|---|---|---|
|  | *less* | *same* | *more* | *less* | *same* | *more* |
| **Proportion of study represented by:** | | | | | | |
| Short courses (open programs) | 2 | 12 | 5 | 1 | 10 | 6 |
| Short courses (tailored programs) | 4 | 8 | 8 | 2 | 10 | 14 |
| Openings courses | 4 | 8 | 10 | 3 | 8 | 7 |
| Level 1 courses | 4 | 7 | 10 | 1 | 9 | 9 |
| Level 2 courses | 1 | 5 | 4 | 1 | 5 | 4 |
| Level 3 courses | 1 | 6 | 5 | 1 | 4 | 6 |
| MA/MSc courses | 0 | 4 | 14 | 0 | 6 | 12 |
| **Proportion of self-financing students** | 2 | 12 | 8 | 3 | 13 | 7 |
| **Proportion of overseas students** | 1 | 5 | 17 | 0 | 12 | 11 |

Data based on questionnaire replies from 23 Open University regions

[The table is an adaptation by the present author of the table on p. 11 of *Student Toolkit 3*]

COMMENTS

The heading could be expressed more clearly. The table is actually about the number of OU regions who run different types of course, and how they perceive changes in proportions of students studying each type of course, but this is not mentioned in the heading. In giving the source of the data used in the table, questionnaire *replies* are mentioned, but there is no mention of how many regions were *approached*, and where geographically, these regions are. So it is not possible to say how typical the results obtained are of the OU in general.

It is not clear what the column headings refer to either. The table includes the heading, 'Proportion of study represented by', followed by various row headings such as 'MA/MSc courses'. "Proportion" in this context usually means percentage, although the label "%" or "percentage" is not included in any of the column or row headings. But none of the row totals sum to more than 23, the number of questionnaire responses on which the table is based. So the heading should perhaps be, 'Number of OU regions who think their student numbers will increase/ will not change/ will decrease'.

Notice also that when the individual row entries for some rows in each column are summed, the result is not the same. For example, the numbers in the first row headed, 'Short courses (open programs)' sum to 19 in the 'Now compared with 2012' column, but to 17 in the 'Projection for 2016 compared with now' column. This means that there are data missing, probably because not all respondents replied to all the questions. Again, this should have been pointed out in the table.

A further point is that the table has a generally 'cramped' appearance: greater spacing between 'MA/MSc courses" and the next main row heading, 'Proportion of self-financing students', and between the latter and 'Proportion of overseas students' would have helped to reduce the table's cluttered appearance.

Alluded to previously in some instances in writing more generally, sometimes in assignments a table is included which is *not* referred to in the text, or is referred to only right at the end, or is hardly discussed at all. It is necessary at least to *mention the table in the text* and as far as possible to *keep the table and any text discussing it together* to avoid confusion (*i.e.* as to what relevance the table has to the ideas in the text- i.e., why it is there) and to-ing and fro-ing for ease of reference.

An example of a well-constructed table is shown below, which is part of the assignment answer on p. 114:

### TABLE SHOWING A RECIPE FOR OAT BISCUITS

| INGREDIENT | IMPERIAL QUANTITY/oz | METRIC QUANTITY/g |
|---|---|---|
| **flour** | 2 | 50 |
| **butter** | 2 | 50 |
| **rolled oats** | 7 | 175 |
| **sugar** | 3 | 7 |

*Source:* Brand *et al* (2005), p. 12

As with graphs, the data *source* should given with the table, as shown in the above table.

A more complex table is shown in an assignment (ECA) for S292. Note how tables can be used, as in this instance, to summarize a lot of information in a relative small space, though as mentioned above, this table probably attempts to summarize too much information:

TABLE 2 SHOWING THE STAGES IN STONE TOOL MANUFACTURE AND USE OVER TIME

| TIME PERIOD | STAGE | TRADITION | Skill required in Production | Refinement to basic tool forms | TOOL TECHNOLOGY Tool Types and Use | Conceptualization of End Product | Innovation During Period |
|---|---|---|---|---|---|---|---|
| Uncertain | 1 | | Odd stones picked up | – | Stones used for hammers/weapons. | – | – |
| * | 2 | | Stones/sharp fragments picked up or shattered | | Hammers/choppers used for breaking bones/ cutting off meat from carcasses. | | |
| 2.6 mya | 3 | Olduvan | Percussion technique; directed blows on selected/rounded cores, removing flakes from either side on one face. Skill increasing over period ending with small bifaces and larger flakes. | – | Rounded hammer stones, crude flakes, "6 unifacial types with chopping, slicing and scraping functions on bone, meat, hide and wood. | Not manufactured for use as specific tools but flakes afterward found useful for a given function. However, flakes struck off to leave surface suitable for further flakes. | Long term continuity of tool types and technique. |
| 1.5 mya | 4 | Acheulean | Larger cores prepared then flakes repeatedly chipped off from both faces to produce different tool types. Detail of preparation increasing over period. | Some later examples more finely hewn. | "10 more massive bifacial types including hand axes, picks and cleavers. Use more effectively for butchering and other tasks. | Regular production to preferred shape. | * |
| 300 kya | 5 | Mousterian | Levallois technique: flakes struck from prepared core shaped into tools. | Finer than typical Acheulean tools. | "40, more specilized, types including serrated flakes and points with spear tangs. Use in more effective hunting weapons for large game and improved butchery | Tools conceived as variations of several standard core shapes. Cutting area maximized to allow reshaping/ sharpening. | Continuation of earlier prepared core technique but quadrupling of tool types |

[cont]

## TABLE 2 SHOWING THE STAGES IN STONE TOOL MANUFACTURE AND USE OVER TIME (CONT)

| TIME PERIOD | STAGE | TRADITION | TOOL TECHNOLOGY | | | | |
|---|---|---|---|---|---|---|---|
| | | | Skill required in Production | Refinement to Basic tool forms | Tool Types and Use | Conceptualization of End Product | Innovation During Period |
| 40 kya | 6 | Chatelperronian | Blade production requiring more time and skill in core preparation. Use of bonefantler points with hammer to detach many blades. ↓ | Still finer tools typical of Upper Paleolithic technologies. ↓ | Additional use of tools/ artifacts of bone, horn, antler. ↓ | | After period of transition between Mousterian and Aurignacian, rapid innovation in techniques and proliferation of tool types and uses. ↓ |
| | | Aurignacian | Increasing complexity of toolmaking techniques. ↓ | Increasing emphasis on aesthetics of products. ↓ | "100 different types over the period. Used in making an increasingly wider and more sophisticated variety of tools, cloth- ing, art, ornaments and hunting weapons. ↓ | Increased stand- ardization and orderliness of production. ↓ | ↓ |
| | | Gravettian | Production of backed blades. ↓ | ↓ | ↓ | ↓ | ↓ |
| | | Solutrean | More elegant tool forms produced by heat treatment. ↓ | ↓ | ↓ | ↓ | ↓ |
| 10 kya | 7 | Magdalenian | Microlith technology: small, fine flakes ↓ | ↓ | Addition of fine arrows, spear thrower heads ↓ | ↓ | ↓ |

Footnotes

† Proposed tentatively, based on chimpanzee behaviour*, **, ***.
‡‡ * * , based on comparisons of chimp and australopithecine brain size/organization, and indications in several studies in the literature*, **.

(1) Stages 3-7 are based on Clark's classification (see Lewin (1989), p.130).
(2) The time periods are based on Lewin, p. 131 the dates given are of the first appearance of the corresponding stage.
(3) The information in Table 2 is based on Lewin (1993)¹, Handprint (1999)ᵃᵇᶜ, Long Foreground ( )ᵃ and Handprint ( )ᵃᵃ.

•

## SUMMARY

To recap, I have tried to show how a number of points arising in working with Graphs, Charts and Tables are of relevance in constructing effective graphs in MST assignments. These points were discussed with reference to the data used in the three types of pictorial representation; their titling and labelling, plotting or tabulation; and on graphs the scales used on the axes, and the drawing of lines through the data points.

In regard to the data presented, it emerged that graphical representation is most appropriate for data where a relationship between a dependent and an independent variable is depicted. Data in the form in which it is collected, or suitably transformed data (in preference to proportions or differences), should be used. The need for proper referencing of the source was then discussed.

In titling and labelling graphs, charts and tables, the importance of a clear, concise and suitably placed heading, and the use of appropriate descriptors for its axes in graphs and charts or column/row headings in tables- maximizing readability which should be the aim in all forms of pictorial representation- were emphasized.

Coming to the question of axis scales, the discussion suggested that axes should ideally be scaled using intervals of 1 or multiples of 10 (2's and 5's may also be used, if necessary). This choice of scale intervals facilitates the plotting of points and the reading of values from the graph. Examples showed that graphs and charts with scales which do not start at zero tend to exaggerate trends in the data. However, it was argued that the use of non-zero starting values may be desirable where data values are confined to a range remote from zero.

In plotting a graph, examples of data were discussed where it was appropriate to join data points with a line-for example certain types of continuous data- and others, notably discrete data, where it was not. An example showed that even with continuous data, joining data points with a line may not be valid for some data sets.

The pitfalls of extrapolating line graphs, i.e., extending them to pass through regions outside the range of the data collected, were also considered.

The section on tables gave examples of poorly and well-constructed tables, showing the need for a heading which adequately describes what the table is about; clear column labelling, and the type data appearing in the table should be defined. The aim is to all essential information 'in a nutshell', with no more than 8 columns and preferable all on one page. Finally, table entries and totals should be checked for agreement numerically.

Several points were made in regard to the referencing and proper use of figures in the accompanying text which again generalize to all three modes of pictorial representation.

In Chapter 6, I go on to describe examples of other types of chart and diagrams, which figure prominently in many answers to 'biological' type essay MST TMA questions.

# CHAPTER 6: OTHER CHARTS AND DIAGRAMS

## CHARTS

This is the following part of the biological question discussed in Appendix 1c above:

> "Draw a food web prior to the onset of myxomatosis, showing the appropriate relationships."

ANSWER:

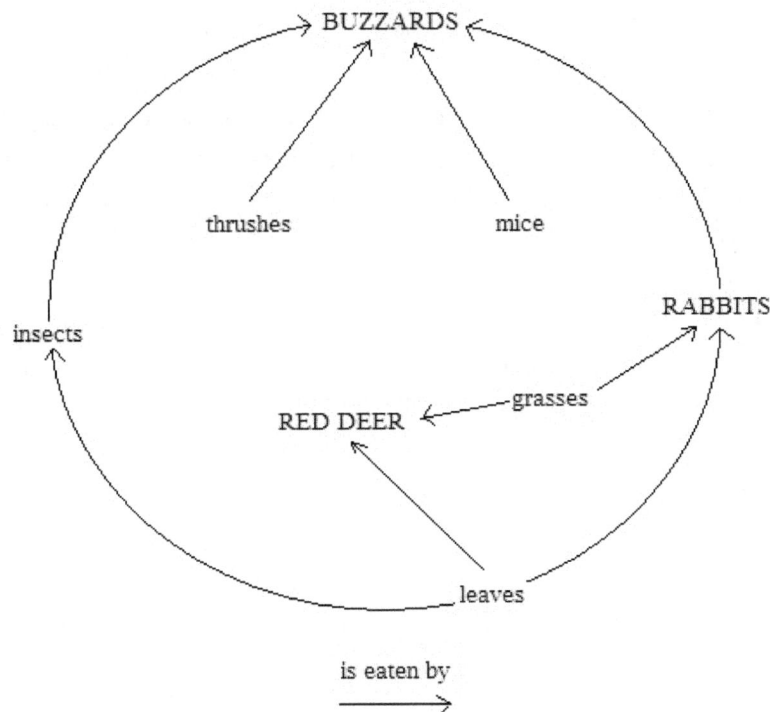

A food web prior to the onset of myxomatosis, showing the appropriate relationships.

Notice how the labelling of the chart is clear and well-spaced, with plenty of room allocated to the arrows representing the feeding relationships between them (as briefly explained by the 'legend' underneath). The main animal species which feature in the question (rabbits, buzzards and deer) are capitalized to highlight them. Finally, what the purpose of the chart and what it is designed to show is written concisely in its title.

The food web shown below is more complex and needed to be carefully organised to make the relationships between the various species clear. This required full use of the space available on an A4 sheet. One colour was used for the relationships in the food web, another for the species, highlighted this time in boxes. The central box was especially emphasized in red since this is the habitat of all the species:

Source: Brand et al 2005, p. 17

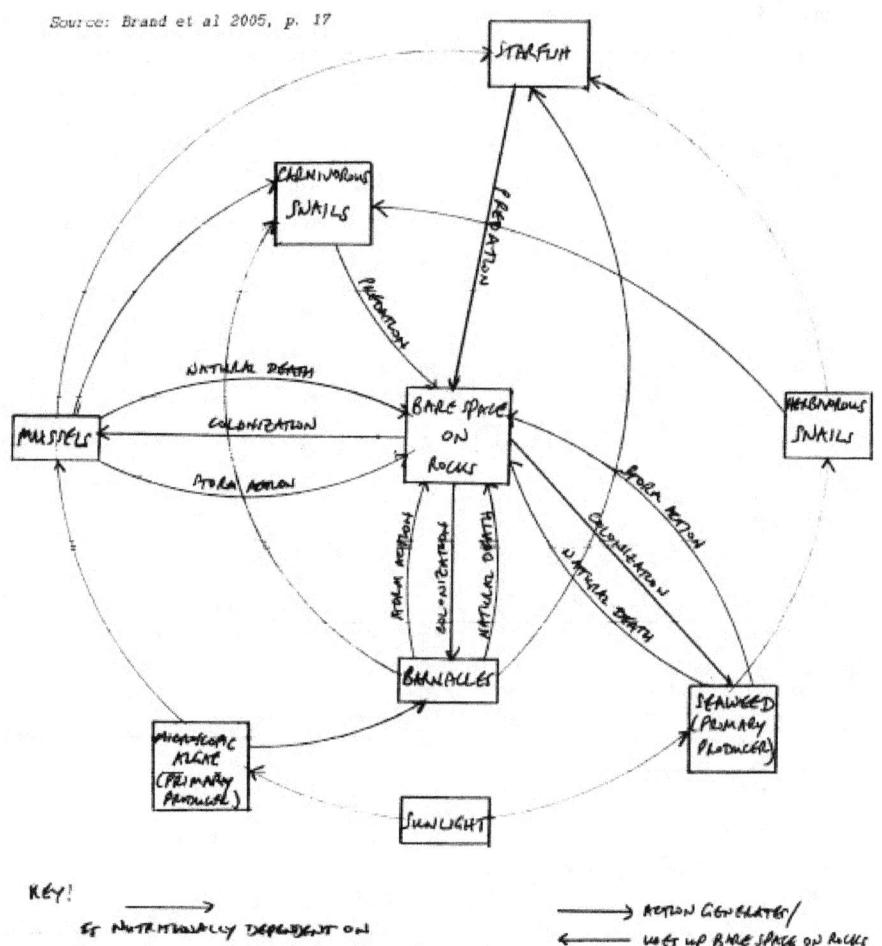

**FIG.3 Showing which species colonize a rocky shore site**

## DIAGRAMS

The following is an example of an answer to a S292 TMA question relating to brain structure requiring a labelled diagram, part of the same question involving the graph of hominoid brain size discussed earlier:

Q.5 (a)

Some indication of when language might have arisen is obtained by comparing the brain organization/structures of extinct hominids with living apes and humans'.

A human-like brain features much more expanded frontal/parietal lobes and smaller occipital lobe (Fig. 2). The lunate sulcus is also used to indicate change from ape-like to more human-like brain organization. Most endocast studies indicate that early *Homo* is the first to show this change².

Broca's area, one of the language centres (Fig. 2) which gives an indication of speech facility, is also more definite in *Homo*.

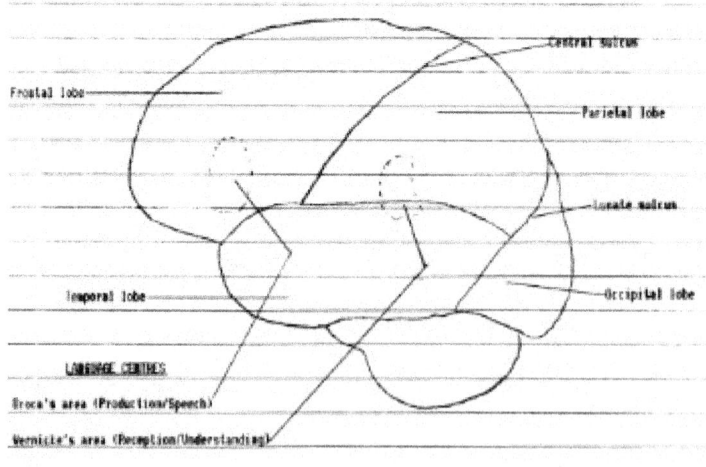

(Adapted from www.aath.tu-dresden.de/below/brain/brainstruc.html)

Fig. 2: simplified sketch of the human brain (left hemisphere) showing the language centres and organization

Notice that the diagram is given a *title* and is identifiable by giving it a *Figure number* (i.e. Fig.1) from a list of figures given elsewhere in the TMA so that it can be easily located. The *source* of the information in the diagram is also given.

The diagram and features in it are *directly referred to* in several places in the text.

The drawing is made *large enough* for all its important features to be clearly distinguishable and as with graphs and charts its labels are well spaced for easy reading.

- 

Having considered the requirements of essay and mathematical answers to more typical MST assignment questions in regard to their *construction* (chapters 1-2) and *content* (chapters 3-6), the following two chapters will discuss the production of answers to two other kinds of assignment: report-type assignment questions and web-page based assignment questions. Chapter 7 considers report-type assignment questions.

# CHAPTER 7: A REPORT-TYPE ASSIGNMENT QUESTION

QUESTION:

**Biscuit recipe Home-Made or Commercial?**

[A recipe for home-made oat biscuits is:
2 oz flour
2 oz butter
7 oz oats
3 oz sugar]

"In a TV program, recipes sent in by different people were tested for fat content. The program claimed that home-made biscuits contained less fat than commercial ones, which contain 11% of fat by weight.

Write a short report [using the information given. To avoid unnecessary repetition here, this is shown in the first table in the answer] explaining whether or not you agree with this claim. Where you have used an average value, explain your choice of average."
[max 300 words]

# ANSWER:

## Do home-made oat biscuits contain less fat than commercially made biscuits?

The purpose of this short report is to investigate the TV program's answer to this question based on the results obtained in the program.

### The TV Program's Claim

The program claimed that, on average, home-made oat biscuits contain less than the 11% fat of commercially-made oat biscuits

### Method: the recipe and fat content of the biscuits made in the TV program

In the TV program, samples of oat biscuit were made to the recipe in Table 1. These samples were then tested for fat content in the TV program.

Table 1: showing the imperial quanities of ingredients in a recipe for oat biscuits and their practical metric equivalents

| INGREDIENT | IMPERIAL QUANTITY/oz | METRIC QUANTITY/g |
|---|---|---|
| flour | 2 | 50 |
| butter | 2 | 50 |
| rolled oats | 7 | 175 |
| sugar | 3 | 75 |

Source: Brand et. al. (2005), p. 12

P. Trigger 73730911   Y153 TMA 01                                    TMA 01

Q.1(b)

Since the proportions of ingredients are the same in each case, the calculated proportion of fat in oat biscuit will be the same whether the original recipe or the metric equivalent is used (vide Table 1, col. 3, p. 3). The metric equivalent will be used here, since the amounts given in the question are metric quantities.

From Table 1, the total weight of ingedients is:

$(50 + 50 + 175 + 75)$ g = 350 g.

The proportion of butter in this amount is given by:

the weight of butter : the total weight of ingredients = 50 g ÷ 350 g = $1/7$.

So, in 100 g of oat biscuit, there is:

100 g × $1/7$ gram of butter.

$90/100$ ths of this is fat because there is 90 g of fat in 100 g of butter (from the question), which gives:

100 g × $1/7$ × $90/100$ = $90/7$ g, after multiplying out and dividing through by ~~100~~.

Now $90/7$ g is 13 g to the nearest gram.

∴ the expected amount of fat in 100 g of oat biscuit is 13 g to the nearest gram.

### This Report's Disagreement with the Claim

The findings of this report do not support the TV Program's claim.

### Reasons: the findings of this Report

Firstly, mathematical working carried out on the results obtained by the TV program (p. 4), revealed that the average weight of fat contained in oat biscuits made in the program was 12% of biscuit weight. This is greater than the 11% average fat content of commercially made biscuits.

Secondly, the expected fat content of oat biscuits made according to the recipe used in the TV program was found to be 13% of biscuit weight. This is also greater than the 11% average fat content of commerically made oat biscuits.

### Choice of Average

The arithmetic mean was chosen as the average to use. This type of average is used by confectionery manufacturers Also, the term "average" used in the TV program's claim is usually understood to refer to the arithmetic mean
(Allen et. al. (1998), p. 65)

Hence, for the purposes of strict comparison between the average fat content of commercially made oat biscuits and the average fat content of oat biscuits made in the TV program, the arithmetic mean was used.

### Results obtained by the TV Program

The results are tabulated below:

Table 2: Proportion of Fat in Oat Biscuits made by 20 different people in the TV Program

AMOUNT OF FAT PER 100 g BISCUIT/g

14 10 11 12 13 13 12 12 13 11 13 9 12 13 13 14 12 12 13 10

*Adapted from* Brand et al (2005), p. 12

### Mathematical Working

MEAN FAT CONTENT OF BISCUITS MADE IN THE TV PROGRAM

Using Allen et al (1998), p. 50,

the arithmetic mean weight of fat/100 g of oat biscuit is given by:

the sum of the all the weights of fat/100 g of biscuit ÷ 20

$$= \frac{9 + (2 \times 10) + (3 \times 11) + (6 \times 12) + (7 \times 13) + (2 \times 14)}{20} \text{ g}$$

$$= \frac{9 + 30 + 33 + 72 + 91 + 28}{20} \text{ g}$$

$$= 242/20 \text{ g}$$

$$= 12.1 \text{ g}.$$

Therefore, the arithmetic mean weight of fat/100 g of oat biscuits made in the TV program is 12 g to the nearest gram, *i.e.*, 12% of biscuit weight to the nearest whole per cent.

## COMMENTS

The purpose of the report is stated in the first paragraph, and the question which the report is investigating follows. Typical in reports*, there are definite, headed sections outlining the Method, Mathematical Working, and the Findings of the report.

\* For an example of a report-type answer to a level 2 MST TMA question (MSXR209) see Trigger (2014b)

Note that the units of measurement are given with all quantities, including the result of any calculation performed with them. Also, note that any formula used is briefly, but clearly defined.

Note also that the other two types of average, the mode and median, might have been used instead of the arithmetic mean- a moot point, perhaps. A histogram of the above data looks like this:

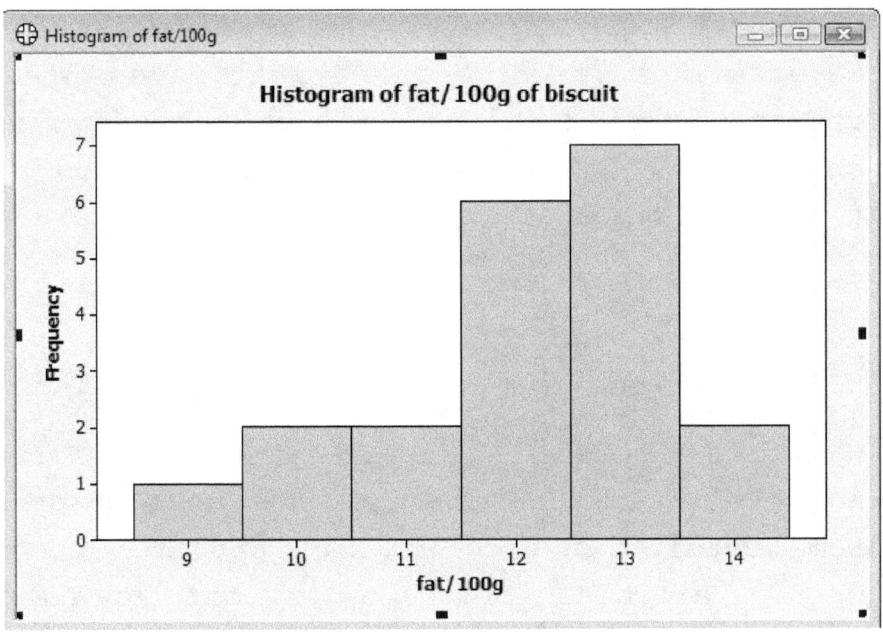

The mean is the best average to choose where there is a symmetrical spread of values about a central value. But here there are more high values than low ones (the distribution is 'right skewed'). The mode is an average which gives the most typical value in the data set, but because the data set here is loaded with higher values, in fact the mode gives a high average of 13g, even though there are almost as many 12g's as 13g's (but note that this figure is the same as that *expected* from the recipe), compared with the arithmetic mean which is 12.1g. So the median, which is the arithmetic mean of the data arranged in ranks and therefore takes account of this, probably gives the best average, and in fact is 12g. Nevertheless, regardless of which average is chosen, the result is higher than the commercial 11g, so the conclusion stated in the above report is still valid in any case.

•

Having discussed report-type assignment answers in this chapter, in the next chapter some detailed advice will be provided about writing articles for the www as part of a MST assignment, a central feature of which is the example given.

# CHAPTER 8: WRITING ASSIGNMENT ARTICLES FOR THE WEB

This chapter makes use of information from OU published sources giving advice about the main aspects of writing for the web, including planning with a clear purpose and an audience in mind; producing an explanation; choosing reliable information from web sources; good web design and publishing on the web. The points raised are further developed and illustrated in an example web-based article assignment question.

## WRITING FOR THE WEB

### Producing a Plan
(abridged from the Open University's T171 Website)

- know your audience
- identify the purpose

#### The audience

Who is your intended audience and what do you know about them? Is it a large audience of strangers? Friends? Colleagues? Are they experts in the subject, reasonably informed or a more general audience?

The answers to these questions should determine the language and level of explanation you use.

#### The Purpose

Know your purpose and make your explanation appropriate to that purpose. For example, if you are trying to persuade your audience that your personal view is correct, a form of explanation expressing opinion is appropriate. Academic writing should be more factual in both its presentation and in its use of supporting evidence...

Having produced a plan the next stage is to consider the structure of the article itself. This involves making clear what the purpose of the article is.

For example, if the purpose of the web-page is to describe or explain something (an example in given in the panel below), the advice given in this extract is apposite:

> **Producing an Explanation**
> (abridged from the Open University's T171 Website)
>
> My Oxford Dictionary defines *explain* as the following;
>
> > make **clear** or **intelligible** with detailed **information**
>
> In formal writing, you will make your arguments in a clear, structured manner, providing detailed information to back up you statements. But an explanation can take many forms, depending on *what* you are trying to explain and *who* you are explaining it to. To illustrate;
>
> > Julie is an IT manager in a large corporation and has been to a special preview of the new Intel microprocessor. The next day she gives a presentation of what she has learnt to her expert colleagues. The explanation is liable to be *technical* and somewhat *objective*, i.e., it will not be influenced by her personal opinions.
> >
> > When Julie gets home, her 10 year-old son wants to know about the new microprocessor which he has seen advertised on TV and whether they will get one for their home PC. This time, Julie will use fewer technical terms, simplify her explanation and concentrate more on benefits to the user.
> >
> > Now imagine that Julie goes out later with some of the same work colleagues she had given her earlier presentation to. During the evening they talk about films and Julie explains why Casablanca is her favourite film. This time, her explanation will be informal and entirely *subjective*, i.e., based on her personal opinions...

Having clearly stated the purpose, the following paragraph(s) in the article should briefly explain how it will go about achieving it. In doing so in the main section of the article, facts should be clearly separated from theories or opinions about them. Likely, the article will draw on other information sources from the web, and the following extract offers advice in choosing reliable sources:

# VETTING MATERIAL FROM THE WEB FOR USE IN MST ASSIGNMENTS

Notes on 'The Principles of Clear Thinking' in Appendix2 'On Clear Thinking' in T171 Preparatory Activities, OU

### INTRODUCTION

Web material contains contradictory views/dubious information. The following aspects should be considered in assessing its validity and reliability;

- ## Context

  *Perspective:*

  What is known about the author(s)?

  What is the author trying to accomplish by putting his material on the Web?

  What does the author base his views on?

  To what extent is he an expert?

  Are ideas kept within the limits to which they apply or are they extended beyond their limits?

  How widely are the author's views shared and by whom?

  What is the author's approach (usually in the first paragraph)

  *Audience*

  What audience is addressed?

- ## Facts

  Facts are ideas which are virtually universally agreed / generally accepted by experts in the field.

> **CONCLUSION**
>
> Help to clarify contradictory/incomplete web info and summarise findings by:
>
> Trying to separate observations from theories and explanations about those observations.
>
> Are events and observations or theories about them reported?
>
> Have observations been made accurately and reported coherently?
>
> The perspective of the author: note any values which may colour what he is says.
>
> ..............................................................................................

In Chapter 1, the *broad structuring* of answers to MST assignment questions in terms of giving a piece of writing a definite beginning, middle and end, was considered. For web articles, this translates into an *Introduction* stating the article's purpose and approach, and a *Main Section* of argument or information in support; a *Summary* and *Conclusion* drawing together the key points made and the inferences drawn from them should then follow.

The next stage is to design the web article in terms of textual formatting, choice and layout of graphics and use of html, which are all aspects of *Good Web Design*:

### Good Web Design
(abridged from the Open University's T171 website)

**Most of the following are easy to do with HTML editors;**

- simple tables to put text or illustrations into columns
- colour for backgrounds, text and cells of tables
- different typefaces, sizes and styles of text
- illustrations to add interest to a page

**Some principles of good Web page design**

- **Responsiveness**

  Make sure your page does not take long to open, otherwise potential readers will not bother to wait and repeat visits will be discouraged.

- **Present a clear and simple structure**

  Make the content and purpose of your Web page obvious- do not include too many bits and pieces or animations which distract the reader.

- **Use colour to enhance not distract**

  Colour can clearly separate different parts of the page, and effective use of coloured text can highlight or distinguish different features according to their purpose. Make sure there is a good contrast between text and background colour. However, a mix of strong, garish colours makes reading more difficult.

- **Provide clear and consistent navigation tools**

  Create a simple table of contents linked to the various sections of your Web page, so that readers can easily and quickly find their way around it. Use a consistent colour text/graphics scheme for your links...

Having produced a first draft the writer should critically review, revise and redraft the web page given the general advice for writing answers to assignment questions in Chapter 2, but also making any necessary alterations to the layout, spacing and other aspects of general design. The appearance of the page when opened on the web, and the functioning of links will also need to be checked.

The final stage is to publish the finished article on the web:

## PUBLISHING YOUR ASSIGNMENT ON THE WEB

### Publishing a Web Page using Microsoft FrontPage Editor

- Connect to the Internet
- Open your Web page in FrontPage Editor
- Click on 'Save As' on the File menu
- In the dialogue box which opens, type in the address of your Web space ( ask your ISP if you don't know what this is ) including the filename of your Web page
- Click ok, then enter your user id and password in the new dialogue box which opens to upload your Web page and any images

The various stages producing a web-based article in answer to an assignment question are further developed and illustrated in the example below, in answer to the TMA question:

> "Design a web page giving advice on how to write effective articles for the Web"

# ANSWER

## Authoring for the Web

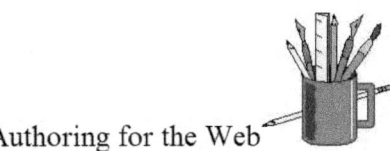

by Peter Trigger

**CONTENTS**

Introduction- the purpose and approach of this document.

I. The purpose of your page- what you are trying to achieve and how you go about showing it.

II. Your intended audience- how your audience affects your writing.

III. Controlling the context of what you write- its objectivity, your perspective and approach.

IV. Making use of Web resources- how to use the Web to enhance your writing.

V. Structuring your writing- organising your writing logically.

VI. Designing your Web page- laying out text and graphics.

VII. Looking critically at the result- reviewing your Web page and getting others' comments.

Producing an explanation describes examples of academic and informal writing.
VIII. Publishing your page on the Web.

Summary

**INTRODUCTION**

This document, aimed at would-be authors of Web pages, identifies a number of stages in the process of developing a page for the Web, and discusses issues which need to be taken into account at each stage.

It is the result of the student author's reflection based on a number of resources including the Open University's T171 website material, relevant external websites and personal experience of reading and writing Web pages.

The stages and some criteria for making decisions at each stage are discussed below.

**I. Decide on the purpose of your page**

Determine the question you are trying to answer in your page and keep it in mind as you write, so that what you write is geared to answering that question.

An academic piece of writing, for example a technological report, should be objective: based on facts backed up with evidence. It should use the passive voice.

An informal piece of writing, for example a personal view of a film, can be more subjective: based on opinion and written in the first person.

Producing an explanation describes examples of academic and informal writing.

**II. *Know your intended audience***

Consider who your intended audience is and what you know about them. This will determine the level at which you write.

# Writing OU MST Assignments

If the audience is expert on the topic of your page, you may assume they have knowledge of the concepts and technical terms underpinning your topic.

If it is the intelligent layman, the use of technical terms should be restricted to the essentials with an appropriate level of explanation and examples.

Providing a plan gives additional advice about writing for different kinds of audience.

## III. Identify the context in which your writing applies

| | |
|---|---|
| Perspective Approach | You should make your perspective clear in order to help readers understand your motives and assess the degree of you expertise. Also, state your approach: how you intend to go about achieving your purpose, in the opening paragraphs. |
| Facts or Theories? | State clearly when you are giving your own opinions or those of others, indicating how generally accepted they are, and whether any observations made are first or second-hand. Be careful to separate theories from facts, and to use them within the limits to which they apply. |
| Context | The Open University's Principles of Clear Thinking offers some pointers about how to choose reliable information for inclusion in your Web page. |

## IV. Make use of Web resources for your Web page content

Install links to other Web pages/sites to enhance and extend the ideas in your page. Include links to sites giving examples or counter-examples, or, as has been done in Section VIII on this Web page, detail some personal experience. Make links to external websites as a means of expanding ideas, including another perspective or providing more information. The previous link above is an example of this.

These devices strengthen your argument whilst adding variety and interest, without significantly increasing the length of your Web page.

## V. Structure your Web page

| | |
|---|---|
| Break down page content into logical units which clearly show how your argument progresses towards the attainment of your purpose. These should be recognizable as a statement of your purpose and approach, a main section of argument and information in support, and a summary drawing together the main points you have made and the inferences you have drawn from them. | Purpose and Approach |
| | Main Section |
| | Summary |

How Users Read on the Web is a brief report of a study of the effectiveness of some additional aspects of structure on the success of Web pages.

## VI. Designing a Web page

### (i) Text

Aim to make you page easy to read by spacing out text into short, headed paragraphs. Add interest by changing text style, size and colour, but always with a purpose. For example, bold or italics can be used to emphasize key points whilst adding variety. Headings in larger size fonts help break up information into more digestible 'chunks'.

### (ii) Graphics

Graphics can add further interest, but should also be relevant to the purpose of your page, for example by illustrating, highlighting or exemplifying points you make.

FreeGraphics and FreeCentre are two websites offering free graphics, but be careful not to breach copyright laws.

### (iii) HTML

A basic understanding of HTML can impove your Web page. For example, a knowledge of directory structure in HTML, ALT tags and Web page naming conventions is useful.

HTML and Web page advice further discusses these and other features of HTML.

However, Web authoring software, such Microsoft's FrontPage Editor, which was used to create this Web page, enables good results to be achieved without detailed knowledge of HTML.

These and other aspects are discussed in Good Web Design.

## VII. Reviewing and piloting your completed Web page

### (i) Content

Look critically at your work to determine which parts need redrafting to better achieve your purpose. It may be necessary to go through several redrafts.

A word processor eases this process with its editing, spell checking and word counting facilities.

### (iii) Piloting

Next, show your completed Web page to a small group of people representative of your audience for their comments.

Redraft your page once again.

### (ii) Design

Make any alterations necessary to your spacing of text, style, headings, colour and images to present a well-structured layout, each having a clear function in relation to the overall design and purpose of your page.

### (iv) Operation

Check the appearance of your page in different browsers and the functioning of links.

## VIII. Publishing your page on the Web

Before publishing your page, give it an appropriate title. Choose a filename which reflects its content in lower case letters with no spaces.

Web publishing means saving your Web page remotely on your ISP's computer. You will need your user name and password. Open your page and save it at the address of your web space. Exact details of how to do this vary slightly with ISP, here is an example, but details are obtainable from your own ISP.

Having published your page on the Web, you may wish to make its presence known to other users of the Web by placing it on the index of a search engine. Here is the author's experience of registering this Web page with AltaVista.

[back to Section IV]

## SUMMARY

This document has identified a number of stages involved in creating a page for the Web, and has indicated how consideration of these stages can lead to a more effective Web page. Drawing on these;

- Decide the purpose of your Web page and be closely guided by that purpose as you write
- Identify your audience and pitch the level of your writing accordingly
- Make clear your perspective, your motives, your level of expertise and the approach you will adopt
    - Distinguish between your use of opinions, theories and facts and use them within the limits to which they apply
- Use Web links to illustrate/strengthen your argument
- Give your Web page a definite structure with an introduction containing your purpose, perspective and approach; a main section of argument in support of that purpose; and a conclusion summarising your main points and the inferences which may be drawn from them

- Design your Web page so as to break it up into readable chunks with varied layout, text styles and appropriate use of headings
    - Add attractiveness by using graphics which are relevant to the purpose of your Web page
- Critically examine the argument, layout and operation of your page then redraft it. Ask others representative of your audience for their comments and redraft it again
- Publish your Web page using the address of your Web space

The production of references to others' work in an assignment answer usually follows as a last stage in producing an answer to an assignment question, but appropriate references may be included with the text as writing proceeds. However, the production of a *list* of REFERENCES may be left to the end of the process assignment writing.

In many instances, as with answers to other types of MST TMA question, the information, analysis or argument presented in an answer to a web-page based MST assignment question will need to be backed up by the use of material produced by other authors to allow the reader to check the evidence, as well as acknowledging the source. This will usually require a brief in-text reference sufficient to locate a fuller reference in a separate list. The kinds of sources used may vary from books, web articles, audios & videos, to OU conference messages, each requiring a slightly different method of referencing. Referencing is the subject of the next chapter, where two commonly used systems are described.

# CHAPTER 8: CITING REFERENCES IN ASSIGNMENT ANSWERS

## INTRODUCTION

ASSIGNMENT BOOKLETs may contain useful advice about using references in assignment answers, but in the author's experience this is not always detailed or comprehensive. The use of video and audio evidence is perhaps less common in academic writing than the use of books. Hence the advice to follow here about referencing video and audio CDs makes a useful addition to that given about books, which may already be familiar to some students from previous reading.

There may be occasions when a student wishes to refer to other types of source, notably documents on the World Wide Web (WWW), articles in books or periodicals, or OU conference messages, where appropriate. How might these sources be referenced? Also, what procedure might be followed when it is necessary to refer to several sources with the same author, or to a book with many authors such as *SGSG*, or to an article in a book containing a collection of articles with different authors?

In the following paragraphs, I attempt to answer such questions, with the aid of examples, by consulting three additional sources of information about referencing. Two of these sources are published by the Open University, and the other on the WWW, *SGSG* and *CITING WWW SOURCES*. (Full details of these sources are given in the REFERENCES here). I also point out some similarities and differences in the system of referencing recommended in these sources and OU

ASSIGNMENT BOOKLETs. For comparison, an alternative system of referencing, which is preferred by some authors or is more appropriate for use with some articles [especially Web articles], known as the Number or Numeric System, is discussed briefly. A final purpose is to give a list, with examples, of the different types of reference covered in this account. This list may be used by students as a resource when writing assignment answers.

First, though, the main reasons why referencing is necessary and what is involved in referencing a source, will be considered.

## REFERENCING A SOURCE

### WHY REFERENCING IS NECESSARY

The *CONCISE OXFORD DICTIONARY* defines 'cite' as follows: 'quote [a] passage in a book or author in support of a point of view'. From the above definition, it is apparent that one reason why the citing of references is necessary is *to allow the reader to check the evidence* on which interpretations are made in the text against the source itself, as noted previously. Therefore, a reference should be of sufficient detail to enable the reader to locate the source as quickly and easily as possible.

Another reason, mentioned in some ASSIGNMENT BOOKLETs and *CITING WWW SOURCES*, is *to acknowledge the source* in respect of its author's work.

**ASPECTS OF REFERENCING: IN-TEXT AND REFERENCE LISTING**

Referencing a source in assignment answers involves two main aspects. The first is a reference to the source in the text itself. This is usually, at minimum, a reference to the author's surname or the source's title, together with the year of publication.

The second is a listing of the full details of the source under the heading of REFERENCES after the text. In some instances it may be possible to give a sufficiently detailed reference in the text alone. This is the case here with some ASSIGNMENT BOOKLETs and the *CONCISE OXFORD DICTIONARY*. However, in many cases, this is too unwieldy and interrupts the flow of the textual account, and so fuller details are usually given in the list of references after the text. In a less formal piece of writing, the use of brief in-text references to source titles seems justified, rather than the more formal use of authors' surnames and years of publication. Full details can still be given after the main text. This procedure was adopted in the present account.

..

**REFERENCING BOOKS**

Book references may be of at least 5 different types. A book may be:

(1) written entirely by one author,

(2) written entirely by several authors,

(3) written by an author or authors unknown

and

(4) written or edited by one or more authors, but containing articles written by others.

In addition, reference may be made to:

(5) the work of authors (known as the secondary source) who are themselves citing another source (known as the primary source), in what *TOOLKIT 1* calls 'secondary referencing'.

Each of these types of referencing will be discussed briefly, and examples of in-text referencing and reference listing of each type will be given.

**(1) BOOKS WITH A SINGLE AUTHOR**

The ASSIGNMENT BOOKLETs, *SGSG* and the TOOLKITs agree substantially about how a book with a single author should be cited in the text. This is to be expected, since all three are OU publications and so recommend the same (the Harvard) system of referencing.

It may be that a general reference to a source is sufficient. For example*:

> Bates (2000) argues that reading can have several purposes...

However, where appropriate, more specific page references should be given (*TOOLKIT 1*). E.g.

>    Bates (2000, p. 21) describes 5 steps to better reading...

Under REFERENCES, the advice given in the Y155 ASSIGNMENT BOOKLET is to write the details of the book in the following order: the author's surname and initial, year of publication, the title in italics, and finally the publisher.

*TOOLKIT 1* differs slightly here, since it includes 'place of publication' before the name of the publisher.
(For completeness, I have followed this procedure here).
Accordingly, the Bates (2000) reference might appear in the list of references as:

Bates, D. (2000) STUDENT TOOLKIT 4 READING AND NOTE TAKING, Milton Keynes, The Open University

(N.B. Notepad had no facility to produce the required italics and so capitals or quotes were used in the original article instead, but under the system described, italics would be preferable).

Where a specific section of the source has been used, page references are also appropriate under REFERENCES. This enables the interested reader to access the relevant source material more quickly. For example, in writing this account, I only used information from *TOOLKIT 1* in the 'Referencing' section on pp. 34-5. Consequently, I cited the full reference as follows:

Johnson, M. and Goodwin, V. (1999) STUDENT TOOLKIT 1 THE EFFECTIVE USE OF ENGLISH, Milton Keynes, The Open University, pp. 34-5

### (2) BOOKS WITH MORE THAN ONE AUTHOR

Where there are several authors, particularly where there are more than three (SGSG for example has four), it may be too unwieldy to refer to each by name in the text. A practice which appears to be common in academic writing is to refer to the first author mentioned on the title page of the book, followed by the phrase 'et al.' (meaning 'and others'). For example:

> Northedge et al. (1997) suggest that diagrams can be misleading...

In the list of references, the full reference to the book might be:

Northedge, A., Thomas, J. Lane, A. and Peasgood, A. (1997) THE SCIENCES GOOD STUDY GUIDE, Milton Keynes, The Open University

### (3) BOOKS CONTAINING ARTICLES WITH OTHER AUTHORS

Where reference is made to an article in a book written or edited by one or more authors, but containing articles written by other authors, the procedure for listing it under REFERENCES is slightly different (*TOOLKIT 1*). For example, an in-text reference might be:

> Green (2002) has investigated the use of statistical diagrams in newspaper articles...

The listing under REFERENCES needs to contain details of the author, date and article title referred to and the beginning and end pages, in addition to the usual details of the book in which the article appears. As book titles are written in italics, the article title is written in Roman type and in quotes. In this case, the complete reference might be:

Green, T. (2002) 'A Survey of Graphs, Charts and Tables in Newspaper Articles', *in* Lovejoy, B. (2003) THE USE AND MISUSE OF STATISICS, Manchester, Addison-Wesley, pp. 170-85

*SGSG* recommends that 'ed' or 'eds' (edited by) be inserted between the last author's initial and the year of publication, where the source is edited.

**(4) BOOKS WITH UNKNOWN AUTHOR(S)**

Here, the title of the book, in italics, could be given in the text (cf. the citing of videocassette references with unknown authors in the *Y155 ASSIGNMENT BOOKLET*), followed by the year of publication (if known). E.g.:

> UNDERSTANDING READING (2003) discusses the skills involved in reading technical texts...

A possible listing under REFERENCES might be:

(2003) UNDERSTANDING READING, McGraw-Hill

However, the reader should be given enough information about the source to enable him or her to locate it. If this is made very difficult, because insufficient information is known to the citing author, the worth of the reference as a source of evidence is degraded.

**(5) SECONDARY REFERENCING**

The work of authors who are themselves referring to another source may be cited in the text by including, in parenthesis, the author of the secondary text. E.g.

> Smith and Peters (1999) (cited in Wilson 2001), note that the display of discrete data on a line graph is invalid...

Under REFERENCES, it is necessary to give full details of both sources:

Smith, J, and Peters, M. (1999) THE PRESENTATION OF
    STATISTICAL DATA, Oxford, Unwin

Wilson, G. (2001) DISPLAYING STATISTICS, Surrey,
    Hamilton Press

**CITING MORE THAN ONE SOURCE WITH THE SAME AUTHOR**

Where different titles published in different years by the same author are cited in several places in the text, the in-text referencing procedure is identical to that used for different authors. This is because these sources are distinguished by their differing years of publication. For example, a book written by Fishbourne in 1999, referred to in one part of the text, and another work by the same author written in 2001 referred to as Fishbourne (2001) in another part of the text, can be readily distinguished in the REFERENCES list by their different years of publication.

Where more than one title with the same author is referred to at the same time, the procedure recommended in TOOLKIT 1 is to quote the author's surname once only, together with the year of publication of each title. For example:

>Fishbourne (1999, 2001) has advanced the view that...

Where more than one title by the same author published in the same year is referred to, TOOLKIT 1 recommends that 'a', 'b', etc. be appended to the year of publication of each source [as is done in this book in previous chapters]. E.g.:

>Fishbourne (1999a) has shown that the use of collected data to plot graphs is more satisfactory than data converted to proportions...

and

>Fishbourne (1999b) recommends the following procedure when deciding which scales to use on the axes...

The procedure for listing the source details under REFERENCES in this case is similar to that involved in referencing sources with different authors.

However, the author's name only need only be written once. All following titles by the same author can be preceded by a string of dashes instead. For example:

Fishbourne, N. (1999a) STATISTICS, Oxford, OUP

--------------- (1999b) MORE STATISTICS, Oxford, OUP

Much of what has been said above about single source, multiple author referencing and single author, multiple source referencing in relation to books, is applicable also to the other types of source discussed below. Consequently, only the differences in basic single author, single source referencing between these other other types of source and books will be discussed.

**REFERENCING PERIODICALS**

Periodical and journal articles are in-text referenced in the same way as books. However, under REFERENCES they should be listed by enclosing their titles in quotes, according to *SGSG* and *TOOLKIT 1*. Whereas the title of the journal or periodical is printed in italics, as in the case of books, article titles are printed in Roman type. An example given in SGSG is:

Collee, G. (1989) 'Inside Science: Food poisoning',
    NEW SCIENTIST, 21, pp. 56-7

**REFERENCING VIDEOs AND AUDIOs**

These are unusual in that the sources are not textual, and the authors may be unknown. The Y155 ASSIGNMENT BOOKLET gives examples of in-text and reference listing for each of these types of source. A point worth adding is in relation to the referencing of particular sections of an audio or video.

One procedure in keeping with the referencing of books is to give the subtitle of the section concerned in Roman type and in quotes, with a frame or part reference at the end. For example, having used the Y155 video as a source of advice in writing assignments, a student might wish to refer to the section of the tape concerned, as follows:

> 'Assignments', OPENINGS RETURNING TO STUDY; YOU CAN DO IT! (2000) video, The Open University, frames 50-80.

Specific parts of an audio disc might be referenced similarly.

## REFERENCING WWW DOCUMENTS

Referencing a WWW document is basically similar to the referencing of printed articles. The WWW document CITING WWW SOURCES points to three differences, though. Firstly, the location of the document (i.e. the address of the website) on the WWW needs to be added to the reference details. Secondly, CITING WWW SOURCES mentions the need to give the author's name only in the text. Then, and thirdly, in the full reference under REFERENCES, the date when the website was visited is included, rather than the year of publication. For example, using this system, an in-text reference might be:

> Graphs are best suited in showing the relation-ship between a dependent and an independent variable (Rafferty).

The full reference under REFERENCES might be:

Rafferty, S. 'Plotting Graphs', HELP WITH STATISTICS,
    http://www.university.edu/graphs.htm
    (28/11/04)

It seems reasonable to suggest, again in keeping with the in-text referencing of book sources, that the author's name be put at the beginning of the reference, together with the year of publication, if known. Under REFERENCES, the year of publication could be distinguished from the date when the website was visited by adding 'accessed:' to the latter:

(in-text)

> Rafferty (2000) suggests that graphs are best suited in showing the relationship between a dependent and an independent variable...

(under REFERENCES)

> Rafferty, S. (2000) 'Plotting Graphs', HELP WITH STATISTICS,
> http://www.university.edu/graphs.htm
> (accessed: 28/11/04)

**REFERENCING OU CONFERENCE MESSAGES**

Having seen how to reference a variety of other types of source, referencing OU messages containing relevant material presents little added difficulty. This case is analogous to that where an article in a book containing a collection of such articles, is referenced. In citing a conference message, the author of the message replaces the author of article, the subject of the message replaces the title of the article, and the name of the conference replaces the title of the book. In this case, of course, the *conference* has no author.

It may be appropriate, as conference messages are usually less formal affairs, to refer briefly to the author and conference in the text, giving fuller details, including the date of the message, under REFERENCES. For example:

> A. Jones in a message to the Y155 conference discusses another method of solving the herb garden problem...

The associated full reference under REFERENCES might be:

Jones, P. 'The Herb Bed Problem: finding a formula', Y155 OPEN UNIVERSITY FIRSTCLASS CONFERENCE message, 28/11/14

**THE ORDERING AND LAYOUT OF THE LIST OF REFERENCES**

An alphabetical listing of references by first author's surname is a suitable scheme under the system described, as used in this book. Certainly, where there are more than a few references, this procedure facilitates a quick location of a particular reference in the list. For this reason also, where the details of a reference occupy more than one line, each subsequent line might begin in from the margin just beyond the leading author's surname. This makes the listing by author more conspicuous and therefore easier to use.

## THE NUMERIC SYSTEM OF REFERENCING

Using this system, numbers instead of authors and dates are used to cite sources in-text (*vide* (=see) *GUIDE TO CITING REFERENCES -NUMERIC SYSTEM*). This may be done using numbers in parenthesis or superscript (*vide ARRANGING & CITING REFERENCES*).

For example:

> However, diagrams can be misleading (1)...

or

> A pie chart may be more appropriate for data in the form of proportions[2]...

The superscript form breaks up the text less than the parenthetical form. However, the parenthetical form is more convenient if it is desired to give in-text page references (and *in a web page-based piece of writing*, is easier to click on [because its font size is bigger] as a link to its place in a list of REFERENCES), although these can still be given in the list of REFERENCES if the superscript form is preferred.

In the list of REFERENCES, instead of placing sources in alphabetical order by author, sources are numbered consecutively in the order they are cited in the text. The above two example in-text references above might appear in the reference list as:

> 1. Northedge, A., Thomas, J. Lane, A. and Peasgood, A.(1997) THE SCIENCES GOOD STUDY GUIDE, Milton Keynes, The Open University

2. Gilmartin, K. and Rex, K. (1999) STUDENT TOOLKIT 6: MORE CHARTS, GRAPHS AND TABLES, Milton Keynes, The Open University.

An example of the use of numeric referencing in a S292 assignment answer is shown below:

**REFERENCES**

(1) Lewin, R. (1998) *Human Evolution*, ...
(2) Lewin, p.192
(3) Evans, L. (2002) 'Australopithecus' in Human Evolution. (http://www.eolss.com/ukbrany_austral.htm)
(4) Lewin, p.80
(5) Goodall, J. (2001) 'Chimpanzees: Hunting' in The Jane Goodall Institute (http://www.janegoodall.org/chimps/hunting.htm)
(6) Lewin, p.57
(7) BBC (2001) 'Ape brains show linguistic Promise. (http://news.bbc.co.uk/1/hi/sci/tech/1680651.stm) 28/11/2001
(8) Goodall, J. (2001) 'Chimpanzees: Communication' in The Jane Goodall Institute (http://www.janegoodall.org/chimps/hunting.htm)
(9) Lewin, p.197
(10) Stafford, A. (—) 'Chimpanzee Communication: an insight into the origin of language'. (http://emuseum.mnsu.edu/cultural/language/chimpanzee.html)
(11) BBC (2000) 'Ancestors walked on Knuckles'. (http://news.bbc.co.uk/1/hi/sci/tech/687341.stm) 22/3/2000
(12) Lewin, p. 108
(13) Lewin, pp.75-6
(14) Martini, P. (1998) 'Australopithecus garhi'. (http://www.rateillo.cc.ca.us/~cromith/garhi.htm)
(15) Schuster, A. (1999) 'New Species Found'. Archgeology Online News. (http://www.archaeology.org/online/news/human.html)
(16) Chang, K. (1999) 'New human ancestor?'. ABC News. (http://www.abcnews.go.com/sections/science/DailyNews/hominid990422.html)
(17) Ash, P. (2002) 'Skull of the Week 5'. S292 OUFC Conference
(18) Kreger, C. (2000) 'Australopithecus afroanus'. (http://www.modernhumanorigins.com/africanus.html)
(19) Handprint (1997) 'The hominid brain'. (http://www.handprint.com/LS/ANC/brain.html)
(20) BBC (1999) 'A Taste for Meat'. (http://news.bbc.co.uk/1/hi/sci/tech/255725.stm)
(21) — (—) 'Australopithecus'. (http://www.humboldt.edu/~mrc1/australo.shtml)
(22) Kreger, C. (2000) 'Australopithecus-paranthropus robustus'. (http://www.modernhumanorigins.com/robustus.html)
(23) BBC (2001) 'Apeman ate termites'. (http://news.bbc.co.uk/1/hi/sci/tech/1119359.stm) 16/1/2001
(24) Lewin, p.119
(25) Evans, L. (2002) 'Homo habilis and Homo erectus' in Human Evolution. (http://www.eolss.com/hollem/hu_habilis.htm)
(26) BBB (1999) 'Ancient Tool Factory uncovered'. (http://news.bbc.co.uk/1/hi/sci/tech/336555.stm) 6/5/1999
(27) Handprint (1999) 'Hominid tools'. (http://www.handprint.com/LS/ANC/stones.html)
(28) Lewin, p.143
(29) Gabunia, L. et al (2000) 'Earliest Pleistocene cranial remains from Dmanisi, Republic of Georgia' *Science*, vol. 288, pp. 1019-1025. (http://www.math.si.edu/anthro/humanorigins/whatshot/2000/wh20004.html)
(30) Lewin, p.103
(31) Lewin, p.196
(32) Lewin, p.146
(33) McCrone, J. (2000) 'The discovery of fire' in *Going Inside*. (www.btinternet.com/~neuronaut/webtwo_features_fire.htm)
(34) Long Foreground, (—) 'Homo Erectus'. (http://www.wsu.edu:8001/vws/gened/learn-modules/top_longfor/timeline/erectus/erectus-a.html)
(35) Kreger, D. (2000) 'Homo heidelbergensis'. (http://www.modernhumanorigins.com/heidelbergensis.html)
(36) Filing Cabinet, S292. 'H.Hedelbergensis'
(37) Lewin, p.186
(38) Handprint (—) 'Homo Hedelbergensis'. (http://www.handprint.com/LS/ANC/homofs.html)
(39) Kreger, D. (2000) 'Homo neanderthalensis'. (http://www.modernhumanorigins.com/heidelbergensis.html)
(40) Lewin, p.156
(41) BBC (2002) 'Neanderthals used glue to make tools'. (http://news.bbc.co.uk/1/hi/sci/tech/1759583.stm) 19/1/2002
(42) BBC (2000) 'Taste for flesh troubled Neanderthals'. (http://news.bbc.co.uk/1/hi/sci/tech/787918.stm) 12/6/2000
(43) Lewin, p. 198
(44) BBC (1999) 'Neoderhthals were cannibals' (http://news.bbc.co.uk/1/hi/sci/tech/462048.stm) 1/10/1999
(45) Long Foreground (—) 'Homo sapiens neanderthalensis'. (http://www.wsu.edu:8001/vws/gened/learn-modules/top_longfor/timeline/neander/neander-e.html)
(46) Kreger, D. (2000) 'Homo sapiens'. (http://www.modernhumanorigins.com/sapiens.html)
(47) BBC (2001) 'Early clues to modern humans'. (http://news.bbc.co.uk/1/hi/sci/tech/1642580.stm)
(48) Whitfield, J. (2002) 'Art history doubles'. Nature. (http://www.nature.com/nsu/020107/020107-11.html)
(49) Lewin, p.195
(50) Lewin, p. 183
(51) Long Foreground (—) 'Homo sapiens sapiens'. (http://www.wsu.edu:8001/vws/gened/learn-modules/top_longfor/timeline/h-sapiens-sapiens/h-sapiens-sapiens-a.html)
(52) BBC (2000) 'Cave Paintings may be oldest yet' (http://news.bbc.co.uk/1/hi/sci/tech/1009953.stm)
(53) Lewin p185
(54) Fig site at Abu Hureya, n. Syria d. 11.5-7 kya (Lewin, p. 215)
(55) Lewin, p. 132
(56) Kreger, D. (2000) 'Homo erectus'. (http://www.modernhumanorigins.com/erectus.html)
(57) Kreger, D. (2000) 'Orrorin tugenensis'. (http://www.modernhumanorigins.com/tugenensis.html)
(58) Lewin, p. 193
(59) Lewin, p. 201f
(60) — (—) 'Renaissance social evolution'. (http://members.aol.com/vemjohnsn/soccevo.html)
(61) Lewin, p. 123
(62) Lewin, p. 101
(63) Lewin, p. 133
(64) Lewin, p. 134
(65) Handprint (—) (http://www.handprint.com/LS/ANC/homofs.html)
(66) BBC (2000) 'Fossils may be first Europeans'. (http://news.bbc.co.uk/1/hi/sci/tech/745080.stm) 11/5/2000
(www.math.tu-dresden.de/~below/brain/brainstrus.htm)
(67) — (—) 'Biological Topic 4 Psychology II Brain Structure and Function Overview and Goals'.

## Where different page numbers in the same work need to be cited at several places in the text

*ibid.* (meaning 'in the same work') with the page number(s), can be used in the list of REFERENCES to avoid unnecessary repetition. For example, the following passage refers to another work by Gilmartin et al:

> Bar charts may be used instead of line graphs where there are several independent variables[3]...

This might appear in the list of REFERENCES as follows:

1. Northedge, A., Thomas, J. Lane, A. and Peasgood, A.(1997) THE SCIENCES GOOD STUDY GUIDE, Milton Keynes, The Open University

2. Gilmartin, K. and Rex, K. (1999) STUDENT TOOLKIT 6: MORE CHARTS, GRAPHS AND TABLES, Milton Keynes,The Open University

3. *ibid.*, p. 15

## Referring to a passage already quoted

use *loc. cit.* (meaning "in the passage already quoted). For example:

> Jones (*loc. cit.*) explains how the method may be adapted to similar problems...

## Where the work of an author is referred to at numerous places within the text

use *passim* (meaning "in many places"). For example:

> Peasgood et. al. (1997, *passim*) have described a number of features of good referencing...

## Emphasis in Quotations

Where the stress on the wording of reference source has been altered in order to make a point of the writer's, for example by italicizing, emboldening or underlining words, this should be acknowledged together with the quotation. For example:

> A pie chart may be more appropriate for data in the form of *proportions* [italics added]

....

# ADVANTAGES AND DISADVANTAGES OF THE HARVARD AND NUMERIC SYSTEMS

Each system has its advantages and disadvantages. The use of numbers instead of authors' names and years of publication is neater and interrupts the flow of text less. This is especially so in short texts with relatively many references. It also avoids unnecessary repetition where the same work is cited at several places in the text. In a long list of references where authors may have surnames beginning with the same letter, numbered references are easier to locate.

On the down side, *REFERENCES TO PUBLISHED WORK* points out that no direct indication is given in the *text* as to the author of the source being cited, and that the presence or absence of an author in the REFERENCES list is less easy to ascertain as authors do not appear in alphabetical order. Also, *CITING BIOGRAPHIC REFERENCES* notes that inserting a new reference *necessitates renumbering* of the list of REFERENCES. On balance, it may be said that in general the Harvard system is more versatile, but in web articles or with short texts employing multiple references the greater in-text conciseness of the Numeric system is advantageous.

· · · · ·

**SUMMARY**

After establishing the need for appropriate referencing in MST assignment answers, this account drew on three OU sources and a WWW source to describe, with examples, how references to a variety of different kinds of source could be cited. This included not only books, but articles, periodicals, WWW documents, conference messages and audio/video discs.

The briefer in-text citing of sources and their associated detailed descriptions in a list of references, were discussed. In this listing, alphabetical ordering by author's surname, and suggestions for the appropriate indentation of references,were described.

The following cases were also discussed: referencing of sources with one author, multiple authors and no known authors; referencing of multiple sources with the same author, and cases where a cited author refers to another source (secondary referencing).

The Numeric system was briefly described as an alternative to the Harvard system of referencing and its advantages/disadvantages discussed.

Finally, a summary list of the different types of author/source referencing covered in this account (Harvard system), with examples, has been compiled. This may be used as a quick-to-access resource when citing references in answers to TMA questions. The list is given below.

......

## SUMMARY LIST OF REFERENCING (HARVARD SYSTEM)

**BASIC REFERENCING**

REFERENCE TO A SOURCE WRITTEN ENTIRELY BY ONE AUTHOR

EXAMPLES
(N.B. The capitals shown here in the examples could be replaced by italics)

In-Text:

   Smith (2000) argues that reading can have several purposes...

(with page reference):

   Smith (2000, p. 31) describes skimming as...

Under REFERENCES :

(Books)

   Smith, A. (2000) READING FOR INFORMATION, Oxford, Oxford University Press

(Periodicals)

   Smith, A. (2000) 'Reading for Information', READING TODAY, 33, pp. 43-51

(WWW Documents)

Smith, A. (2000) 'Reading for Information', READING SKILLS, http://www.university.edu/reading.htm (28/11/04)

(Conference message)

> Smith, A. 'Reading for Information', E104 OPEN UNIVERSITY FIRSTCLASS CONFERENCE message, 28/11/04

(Video/audio disk)

> Smith, A. (2000) READING FOR INFORMATION audio CD, Association of Advanced Readers

**REFERENCE TO A SOURCE WRITTEN ENTIRELY BY MORE THAN ONE AUTHOR**

EXAMPLES

In-Text:

> Johnson and Goodwin (1999) note that a sentence needs a subject and a verb...

(Multiple authors)

> Northedge et al. (1997) suggest that diagrams can be misleading...

Under REFERENCES:

> Northedge, A., Thomas, J., Lane, A. and Peasgood, A.(1997)THE SCIENCES GOOD STUDY GUIDE, Milton Keynes, The Open University

(Similar changes to the procedure exemplified under BASIC REFERENCING above are required in referencing other types of source).

## REFERENCE TO SOURCES WHICH ARE CITING OTHER SOURCES

EXAMPLES

In-Text:

>Smith and Peters (1999) (cited in Wilson, 2001), note that discrete data...

Under REFERENCES:
(Quote *both* primary and secondary sources)

>Smith, J. and Peters, M. (1999) THE PRESENTATION OF STATISTICAL DATA, Oxford, Unwin

>Wilson, G. (2001) DISPLAYING STATISTICS, Surrey, McGraw-Hill

## REFERENCE TO A SOURCE WITHIN ANOTHER SOURCE

EXAMPLES

In-Text:

>Green (2002) has investigated the use of statistical diagrams in newspaper articles...

Under REFERENCES:

>Green, T. (2002) 'A Survey of Charts, Graphs and Tables' *in* Lovejoy, B. (2003) THE USE AND MISUSE OF STATISTICS, Manchester, Addison-Wesley, pp.16-31

(add 'ed(s)' after author(s)' surname if edited).

**REFERENCE TO A SOURCE WITH AUTHOR/AUTHORS UNKNOWN**
EXAMPLES

In-Text:

>   MAKING MATHEMATICS ACCESSIBLE (2003) argues that some maths texts are too symbolic...

Under REFERENCES:

(Books-rarely)

>   (2003) MAKING MATHEMATICS ACCESSIBLE, Harrap Press

(Audio/videos)

>   (2003) MAKING MATHEMATICS ACCESSIBLE audio CD, The Education Foundation

(similar changes are required to the BASIC REFERENCING of other types of source).


**REFERENCING MORE THAN ONE SOURCE WITH THE SAME AUTHOR**

EXAMPLES

In-Text:

(Different years of publication, more than one source referred to simultaneously)

>   Fishbourne (1999, 2001) has advanced the view that...

(Same year of publication)

> Fishbourne (1999a) shows that data in the form of proportions...

(Same year, simultaneous reference)

> Fishbourne (1999a, 1999b) shows that percentage data are best displayed...

Under REFERENCES:

> Fishbourne, T. (1999a) STATISTICS, Oxford, OUP
>
> -------------- (1999b) MORE STATISTICS, "

(Similar changes are required to the BASIC REFERENCING of other types of source).

.......

---

§ The example references are for the purposes of illustration only.

.

The following article, as part of an assignment for a level 1 Technology course, (TU120 *Beyond Google*) illustrates some of the points made above.

## CAN WALKING REDUCE STRESS?
By P. Trigger, 1 December 2009

The type of "stress" of interest here is *psychological* stress which tends to detract from a feeling of general well-being. Thus the stressed person may feel tense or unrelaxed, lacking in calm or even unhappy. A general lowering in spirit is also an indicator of stress. Symptoms such as anxiety, irritability and sleeplessness might well accompany stress (Owens).

It is common knowledge that the pressures of daily living - in the workplace, on the roads, at home, in the day-to-day relationships with others -can increase stress. How then might daily living be changed to decrease stress?

The benefit of physical activity in reducing stress is well documented (Department of Health (2004)). For example, Norris, Carroll, and Raymond (2002) in a study of 80 subjects assigned to high or moderate intensity aerobic training or a control group found some self-reported benefits with regard to psychological stress.

Fredericks claims that soothing hormones are produced in the body during sustained exercise, a view also held by Sykes, an academic in the field of Exercise and Nutrition Science. Sykes (2009) says that the effect of exercise is to stimulate increased blood flow and the release of chemical endorphins into the blood stream, which cause a natural calming effect.

So the effect of exercise in reducing stress appears to be well-established and seems to have a scientific basis. But can milder forms of exercise such as walking have a beneficial effect? Owens argues that any form of aerobic exercise, including walking, calms the nerves thereby reducing stress. Although Owens believes that a gentle walking pace can be just as effective as a fast one, given a pleasant, noise-free walking environment, there is other evidence that benefit increases with walking speed and amount. Sykes suggests that:

> ```
> The best way to achieve an increase in
> positive mood appears to be via moderate
> intensity exercise, such as
> ```
> *brisk*
> ```
> walking.
> ```
> (Italics added).

To reduce stress, The World Health Organisation recommends starting off with two walks ("brisk or not") of ten minutes on three days per week in the first week, two walks of 12 minutes on four days per week, building up after 5 weeks to walks of 15 minutes including gentle uphill stretches four days a week.

So, the answer to the question "Can walking reduce stress?" appears to be "Yes" .Walking causes the body to release chemical substances into the bloodstream which produce a calming effect, alleviating stress. It seems that more benefit is obtained by going on regular and frequent walks (on three or more days a week) for at least ten minutes, walking at a brisk pace. The effect may be enhanced if gentle uphill stretches can be encompassed in a walk in pleasant, noise-free surroundings.


REFERENCES

British Heart Foundation *Walking the Way to Health* http://www.whi.org (accessed 31 October, 2009)

Department of Health (2004) *At least five a week: Evidence on the impact of physical activity and its relationship to health. A report from the Chief Medical Officer* London

Frederick, S. *5 Easy Ways to Reduce Stress*
http://www.holistic.com/holistic/learning.nsf/title/5+Easy+Ways+to+Reduce+Stress (accessed 31 October, 2009)

Norris, R., Carroll, D, Cochrane, R. *The effects of physical activity and exercise training on psychological stress and well-being in an adolescent population* Journal of Psychosomatic Research. Vol 36(1), Jan 1992, 55-65.

Owens, D. H. *Walking and Stress Reduction* http://www.cyberparent.com/walks-walking/stress-reduction-benefits.htm (accessed 31 October, 2009)

Sykes K. *Healthy Step*s 'Occupational Health, Sep2009, Vol. 61 Issue 9, pp. 40-43

- 

The next, and final, chapter contains a summary of all the advice given in Chapters 1-9 of this book, which is followed by some conclusions based on it.

# CHAPTER 10: SUMMARY AND CONCLUSION

## SUMMARY

A brief outline overall summary of the main chapters of the book will be given, followed by a more detailed chapter-by-chapter summary. Chapters 1 and 2 were concerned with the written medium and general **construction** of assignment answers to **essay-type** questions, mathematical-type questions and questions involving a mix of the two. Chapters 3 to 6 were concerned with TMA *content*: methods of **presenting information** and *mathematical subject matter*. Chapters 3 and 4 dealt with the *pictorial depiction of information*, Chapters 5 and 6 with mathematical topics. Chapters 7 and 8 then described and discussed answers two further types of MST assignment question- **reports** and **web-based assgnments**. The final chapter dealt with the citing of references.

**Chapter 1** discussed the handwriting and typewriting (word processing and printing out) of assignments. Typewriting gives the neatest and most readable result, but this is more difficult in the case of mathematical assignments where symbols which are not part of the standard QWERTY keyboard have to be included. I described how when I submitted such assignments in wordprocessed form, I was told by one tutor that I was the only one of his students to do so. But then another tutor urged the opposite, as described in Chapter 1. The bottom line is, whichever method is used, and it was explained that both could be used in a given assignment by typing words and adding in special symbols by hand, the answer must be legible and clear. An example assignment answer was shown in which marks were lost because it was not.

**Chapter 2** described one approach to producing MST assignment answers, breaking the process down into stages. The ***first stage*** is to carefully read the question and

>    (a) decide what the question is asking

>    (b) how it is to be answered.

To do this, the advice was to identify key words in the question and discern their meanings.

The ***second stage*** is to construct a plan of the answer in terms of the three aspects of *brainstorming* for ideas to include in the answer; *selecting* the most relevant of these and *sequencing* them in a logical order.

The ***third stage*** is to write an initial draft

of the answer, *broad structuring* the answer to give it a definite **introduction, main section** or body of argument and a **conclusion**, placing ideas in each of these sections based on the outcome of the *sequencing* stage. *Fine structuring*, including paragraph construction and the use of *link words and phrases* in sentences within and between paragraphs; appropriate use of *technical language;* and various devices to *introduce variety,* were all discussed.

The **fourth stage** is to write subsequent drafts, going through each previous draft to improve its logical progression and fluency, changing to more concise wording and excising non-essential or redundant ideas to meet any word count limit imposed by the question within the tolerance allowed by the marking tutor, who might be asked what this is.

The **fifth and final stage** is to check through the answer, correcting spelling mistakes, omissions and grammatical errors, finally reading through the answer once more. I mentioned that I have found it beneficial, time permitting, to return to the answer after two weeks or so, when further improvements might become apparent. Getting others' comments can also be very helpful.

The application of the method was described in in answering example essay and mathematical type assignment questions.

**Chapter 3** provided advice in coping with some basic mathematics topics of substitution in numerical values in and rearrangement of formulae, solving linear equations, S.I. units of measurement, powers of 10, scientific notation, decimal places and significant figures, commonly involved answering entry-level MST assignment questions. The provision of a slightly different approach to explaining an idea and in the working of examples to that presented in course units was seen as one possible benefit in that an alternative approach can sometimes succeed better in getting an idea across to some students depending on their 'learning styles'.

It argued that additional sources of help may also offer further examples on which to practise mathematical skills in helping students to increase their confidence and ability in number work and algebra. It was noted that the general advice given about the use of supplementary material to help establish and reinforce concepts is also apposite in the case of the two other areas of MST (*i.e.* Science and Technology).

**Chapter 4** was concerned with laying out and the sequencing of the logic involved in deriving a mathematical formula describing the number of flower beds in a garden, given its length L and width W, which was determined by counting and inspection. A proof that the derived formula is generally true, of more interest to those students studying level 2 mathematics courses, was included for completeness.

**Chapter 5** offered advice on how to (and how not to) construct MST graphs, charts and tables. Line graphs, pie charts and bar charts were all discussed, with examples. The advice included the use of simple but clear headings; attribution of data source; and the use of appropriate scale, origins and intervals on graphs and charts. In the latter case, where this is not done, examples were given of graphs and charts tending to give a false impression of trends in the data.

Line graphs are used to depict a *relationship between two variables*. Both linear and non-linear graphs were covered. Where such graphs are used for predictive purposes, examples of the consequences of going too far outside the range of data were described.

Several examples of tables were discussed. The first was an example of a poorly conceived table that in its layout was cramped; in which the heading did not adequately describe what the table was about: the column labelling was unclear, the type of data appearing in the table was misrepresented as "proportions" instead of frequencies (numbers of responses), and the representativeness of the data of the population of which they were a sample could not be ascertained, and therefore the generality of the conclusions drawn from the table about the changing patterns of study of OU courses was uncertain.

Subsequent examples featured tables which overcame the above pitfalls, the first a 4 rows x 3 columns table with numeric data, emphasizing simplicity, and featuring a well-spaced and clear layout. The second was a two-part table, showing how a lot of information (but needing to avoid too much), might be summarized in a more complex table occupying several pages, provided the number of columns, with a maximum of around 8, is not exceeded.

**Chapter 6** described examples of other types of chart and diagrams, in this case in answer to 'biological' type essay assignment questions. The first two were 'food webs', the initial example showing a simple circular layout with all the major species properly labelled and well separated on the chart, with arrows clearly showing the feeding relationships between them.

The second chart was more complex, but still succeeds in adequately representing 3 other types of relationship in addition to feeding. Clarity was achieved by careful layout and spacing, the placing of the species in boxes in one colour, and the different relationships between them in others, with a third, more outstanding colour for the habitat common to all the species depicted.

The third example was of a diagram showing a human brain with its language centres, and organization. It was noted how the diagram was made large enough for the various structures and full in-diagram textual outlines of their features to be distinguished from one another, which were clearly labelled. As with all graphs, charts and diagrams, the need to actually refer to the figure from within the text was emphasized.

**Chapter 7** described how to structure the answer to a report-type assignment question. Headed sections outlining the Method, Results, Mathematical Working, and the Findings of the report were described. The need to provide units of measurement with all quantities, including those used in calculations, and to define any formula used

# Writing OU MST Assignments 165

were discussed. There followed a discussion of the 3 kinds of average and choosing an appropriate average from the three for a given data set which often figure in the statistics of answers to MST assignment questions.

**Chapter 8** gave advice, with an example, about writing articles for the www as part of a MST assignment. In *Producing a Plan* it was noted that in constructing a plan the student writer should know their *audience*, *i.e.*, the readers to whom the piece of writing is directed. Also, the piece should have a definite *purpose* and its content should be appropriate to that purpose. *Producing an Explanation* indicated the importance of tailoring an article in terms of presentation, level and use of technical language to match the audience. Clearly these would all be different for one audience of consisting of other MST students with knowledge of MST compared with another of more general readers. The need to separate facts from theories and opinions was emphasized. The extract *Vetting material from the Web for use in MST assignments* offered advice in choosing reliable sources of information for the article.

Having structured the content of the piece of writing in terms of the general advice given in Chapter 2, the next stage is to design the web article in terms of textual formatting, choice and layout of graphics and use of html. To this end, *Good Web Design* focused on some basic facilities offered by HTML editors such as choice of fonts, colours, tables and illustrations and noted the importance of a reasonable web-page opening time, a clear and simple structure, the use of colour to enhance but not distract, and the provision of clear and consistent navigation tools. The final stage is to publish the finished article on the web which was the focus of *Publishing Your Assignment on the Web*. This extract outlined the steps involved in uploading the finished article to a suitable web-site or space.

Finally, the web-based assignment article entitled, *Authoring for the Web* simultaneously provided an additional explanation and a discussion of the various stages in writing a web article previously outlined whilst putting the advice given into practice in the content, structure, and presentation of the web article itself.

Lastly, **Chapter 9** dealt with referencing and the compilation of a REFERENCES list in what is usually the last stage in writing answers to assignment questions. After establishing the need for appropriate referencing in assignment answers, it described, with examples, how references to a variety of different kinds of source could be cited. This included not only books, but articles, periodicals, WWW documents, conference messages and audio/video discs. In-text referencing and the construction and layout of a list under a separate section of REFERENCES, in both *Harvard* and *Numeric* referencing systems, were described, and some advantages and disadvantages of each were outlined.

- 

The final section draws out some general conclusions from all the advice given about writing answers to MST TMA questions.

## CONCLUSION

This book has discussed essay-type, mathematical-type, mixed, 'biological', report-type and web page-based assignments. Though the features of effective answers to each of these types of question differ in detail, certain features of 'good practice' which are common to writing MST TMAs in general emerge from their discussion.

To start with, in producing *any* assignment, whatever medium is used, the writer must keep in mind that it is essentially a *message which is communicated to others*, the marking tutor in particular, and therefore if this communication is to be successful its message must be received. This requires that it is first and foremost clear and *legible*.

Before starting to write the answer to any assignment question, *the question must be read very carefully* and analyzed, picking out its keywords and discerning their precise meanings in order to *closely match the content and approach of the answer to* the requirements of *the question*.

In regard to *construction,* effective answers tend to be well planned. So, the answer to any MST TMA question will benefit from brainstorming, selecting and sequencing, and structuring activities, allocating and organizing ideas in an initial *plan*, as a prelude to writing the first draft. Subsequent redrafting and checking are also desirable aspects in the writing of any well-constructed answer to any MST TMA question.

Information and ideas from other sources, including OU course materials will used, and the main principles of citing in-text referencing and producing a list of references will apply in answers to all types of MST assignment questions.

In regard to *content,* there are two main aspects: (1) *subject matter* content-information, topics and concepts which form the 'stuff' of the knowledge areas of mathematics, science and technology; and (2) *pictorial* content as means of presenting information - graphs, tables, mathematical and 'biological' charts and diagrams.

A firm grasp of the *subject matter* of the assignment in terms of the MST topics and concepts involved is obviously essential if the student is to apply them effectively in answering an assignment question. With this in mind, it was argued that whilst in many cases OU course materials may be sufficient, some students, in some topic areas, may find they need additional help or would benefit from sources supplementary to the OU course materials which may provide an alternative approach and extra examples to improve their understanding and application of ideas in their TMA answers, thereby improving the mark they achieve. First and foremost, this means identifying precisely what these ideas are from the analysis of the question into key words and phrases carried out in making a plan of the assignment answer. Although a number of basic *mathematical* topics and associated concepts were identified which commonly feature in the subject matter of elementary MST course TMAs, the general advice given applies in principle to all MST assignments whether the subject matter is *mathematical*, *scientific* or *technological*.

Answers to MST assignment questions will very likely employ figural presentations of data or other information, and again there are certain features of good practice which apply to graphs, charts, tables and diagrams alike. An essential feature is *readability* and to this end, layouts should contain the minimum in essential information whilst making maximum use of space. Headings should be clear, concise and suitably placed. Actual rather than converted data is preferable for use in figural presentations, and the data source should be made clear and properly attributed and included with the figure itself. And of course, it is important to refer to any figure in the text accompanying it so that the reason for its presence and its purpose are explained to the reader. In addition, the figure should be numbered and listed so that it can be quickly located. Finally, any calculations appearing as summary statistics based on information in the figure should be checked against the data.

. . . . . . . . . .

# REFERENCES AND BIBLIOGRAPHY

ARRANGING AND CITING REFERENCES - A SHORT GUIDE
FOR ENGINEERS
http://www2.ntu.ac.uk/llr/shortcite.htm (19/12/04)

CITING BIBLIOGRAPHIC REFERENCES
http://www.cranfield.ac.uk/cils/library/find/
citing_bibliographic_references.doc.
(19/12/04)

CITING REFERENCES
http://legacy.uwcm.ac.uk/support/libraries/
guides/citing_references.doc
(19/12/04)

CITING REFERENCES
http://www.library.rdg.ac.uk/help/citing/
(19/12/04)

Cooke, H., Allen, B., Holmes, H. and Evans, H. (1998)
'Breakthrough to mathematics, Science and Technology:
Module 4 Study Book Exploring Pattern', Open
University.

Course Information & Advice Centre
The Open University PO BOX 72, Walton Hall, Milton
Keynes, MK7 6ZS

Davy, T. GUIDE TO CITING REFERENCES
- NUMERIC SYSTEM
http://www.libary.stir.ac.uk/
(19/12/04)

Dinchak, M. 'Citing WWW Sources', CITING ELECTRONIC
SOURCES

Fitzpatrick, M., Williams, R. and Peasgood, A.(1999) 'Breakthrough to mathematics, Science and Technology: Module 5 Study Book The Material World', Open University.

FRS (Fisheries Research Services) (2002) *Statistical Fisheries Data for ICES rectangle 44E6 and 45E5 in 2001.Unpublished data. Supplied by FRS, SERAD.*

Gilmartin, K. and Rex, K. (1999) 'Student Toolkit 3: Working with Charts, Graphsa nd Tables', Open University.

Gilmartin, K. Laird, H. and Rex, K. (2001) 'Student Toolkit 7: Maths for Science and Technology', Open University.

------------------------------- 'Student Toolkit 6: More Charts, Graphs and Tables', Open University.

Goodwin, V. and Bishop, J. (1999) 'Student Toolkit 2: Revision and Examinations',Open University.

http://staff.gc.maricopa.edu/~mdinchak/eng101/citing.htm (1/1/00)

Johnson, M. and Goodwin, V. (1999) 'Student Toolkit 1: The Effective Use of English', Open University.

Manning, E. and Houston, M. (2000) 'Student Toolkit 5: Essay and Report Writing Skills', Open University.

Northedge, A., Thomas, J., Lane, A. and Peasgood, A. (1998) 'The Sciences Good Study Guide', Open University.

Talisman Energy UK (2005) *Scoping Document: Beatrice Deomonstrator Windfarm*

Trask, L.(1997) REFERENCES TO PUBLISHED WORK *http://www.informatics.susx.ac.uk/doc/punctuation/node=49.html(accessed 19/12/04)*

Trigger, P. 'Help with Y155 maths topics available in Student Toolkit 7',Y155 Open University FirstClass Conference message 23/11/04

----------- (2013) 'Wind Turbines: Description, Appraisal & Alternatives, Vol. 2' KDP Durham

----------- (2014a) 'Historical Computing Vol. 1: Programming in *BASIC*' KDP Durham

----------- (2014b) 'The Econometrics of Domestic Hot Water Pipe Insulation: An Investigative Analysis', KDP Durham

www.ingramcontent.com/pod-product-compliance
Lightning Source LLC
Chambersburg PA
CBHW051700170526
45167CB00002B/473